7th International Conference on Advanced Materials, Mechanics and Structural Engineering (AMMSE 2020)

IOP Conference Series: Materials Science and Engineering
Volume 1048

Taipei, Taiwan
25 - 27 September 2020

ISBN: 978-1-7138-3307-9
ISSN: 1757-8981

Printed from e-media with permission by:

Curran Associates, Inc.
57 Morehouse Lane
Red Hook, NY 12571

CURRAN ASSOCIATES INC.
proceedings.com

Some format issues inherent in the e-media version may also appear in this print version.

Institute of Physics
Dirac House, Temple Back
Bristol BS1 6BE UK

Phone: 44 1 17 929 7481
Fax: 44 1 17 920 0979

techtracking@iop.org

Additional copies of this publication are available from:

Curran Associates, Inc.
57 Morehouse Lane
Red Hook, NY 12571 USA
Phone: 845-758-0400
Fax: 845-758-2633
Email: curran@proceedings.com
Web: www.proceedings.com

TABLE OF CONTENTS

IOP Conf. Series: Materials Science and Engineering 1048 (2021) 011001 doi:10.1088/1757-899X/1048/1/011001

Preface

The AMMSE 2020 is the 2020 7th International Conference on Advanced Materials, Mechanics and Structural Engineering (7th AMMSE 2020) will be held in Taipei, Taiwan during September 25-27, 2020. But currently, the entire world is struggling against the virulent pandemic COVID-19, so that the AMMSE 2020 conference changed the presentation method from Oral presentation to Remote Video Presentation to help with Global Quarantine Activities.

The Virtual of AMMSE 2020 was held on Incheon National University at Video Conference Meeting Room on September 25-27, 2020. The AMMSE Committee members attended conference at on-site and also authors attended the conference as Video Meeting. The conference program covered Invited Keynote Speech, Oral, and Poster Presentations, they presented their presentation 15 minutes including Q&A time.

The committee of AMMSE 2020 expresses their sincere thanks to all authors for their high-quality research papers and careful presentations. All reviewers are also thanked for their careful comments and advices. Thanks are finally given to IOP Publication as well for producing this volume.

The Organizing Committee of AMMSE 2020

Committee Chair

Prof. Mosbeh Kaloop

Incheon Disaster Prevention Research Center

Incheon National University

IOP Conf. Series: Materials Science and Engineering 1048 (2021) 011001 doi:10.1088/1757-899X/1048/1/011001

Conference Photograph

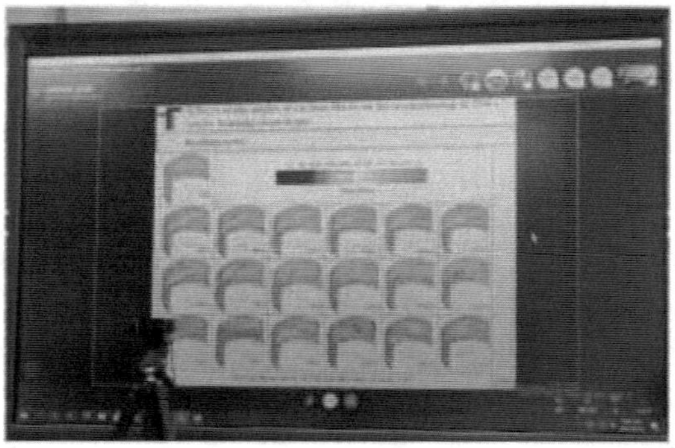

Sponsors

Incheon Disaster Prevention Research Center (IDPRC)

Recently, various efforts to prevent and prepare are vitally needed for prevention of Disasters and calamities. So we understand the necessity for technology of disaster and we built up the Incheon Disaster Prevention Research Center (IDPRC) in Incheon National University (INU) in 1997.

Accordingly, Incheon Disaster Prevention Research Center (IDPRC) in Incheon National University has progressed of research on the prevention of disasters and calamities through the various seminars, conference and lectures. These research could be conducted cooperation with Incheon National University (INU) in various fields' structure, soil, hydraulic and environment.

Incheon Disaster Prevention Research Center (IDPRC) will try to be a leader in the disaster of industry through the various research activities and global conference.

IOP Conf. Series: Materials Science and Engineering 1048 (2021) 011001 doi:10.1088/1757-899X/1048/1/011001

Lists of Committees

Committee name	Organizer name	Organization
Honorary Chair	Mosbeh Kaloop	Incheon National University
Conference General Chairs	Mosbeh Kaloop	Incheon National University
Honorary Chair	Dongkeon Kim	Dong-A University
Technical Committees	J. W. Hu	Incheon National University

Peer review declaration

All papers published in this volume of *IOP Conference Series: Materials Science and Engineering* have been peer reviewed through processes administered by the Editors. Reviews were conducted by expert referees to the professional and scientific standards expected of a proceedings journal published by IOP Publishing.

- **Type of peer review:**
  : Triple-blind (Three of Committee Members)

- **Conference submission management system:**
  : Our conference has Secretary for management on conference papers. So authors send their papers to Conference E-mail, and our secretary has managed it.

- **Number of submissions received:**
  : Especially in this time, only few of papers were submitted cause of COVID-19. The number of submission papers are 78 papers.

- **Number of submissions sent for review:**
  : The submission papers are 50 papers for review.

- **Number of submissions accepted:**
  : The accepted papers are 26 papers.

- **Acceptance Rate (Number of Submissions Accepted / Number of Submissions Received X 100):** 30 %

- **Average number of reviews per paper:**
  : Three times review on per paper.

- **Total number of reviewers involved:**
  : Three of reviewers are took part in the review process.

- **Any additional info on review process:**
  : We send the Review Materials to author through E-mail, so that author can check the three of review comments papers about their papers. This process also gives a last chance to authors for revision.

- **Contact person for queries:**
  : Y.H. Joo (E-mail: ammse2020@ammse-conf.org)

High and Low Temperature Resistance of Phenyl silicone rubber

Yanhua Huang[1,a*], Qiuhong Mu[2,3], Zhengtao Su[1]

[1]Beijing Institute of Aeronautical Materials , Aviation Key Laboratory of Science and Technology on Materials and Application Research for vibration and Noise Reduction, Beijing 100095 , China
[2]Qilu University of Technology (Shandong Academy of Sciences), Advanced Materials Institute, Jinan 250014, China
[3]Shandong Provincial Key Laboratory for Special Silicone-Containing Materials, Jinan 250014, China

E-mail: hyhsdu@163.com

Abstract: Methyl Phenyl and diphenyl silicone rubbers were synthesized from octamethyl cyclotetrasiloxane (D_4), methylphenyl cyclosiloxane, octaphenyl cyclotetrasiloxane, etc. Compared with the commercial methylvinyl silicone rubber, the high and low temperature resistances of methyl vinyl silicone rubber were studied in this paper. The results showed that the introduction of phenyl groups can improve the thermal properties of silicone rubber, and the thermal stability of diphenyl silicone rubber is better than that of methyl phenyl silicone rubber. But there is little difference of mechanical properties in high temperature environment for three silicone rubber. DMTA analysis showed that there were still a small amount of crystals in diphenyl silicone rubber, while methylphenyl silicone rubber was amorphous rubber. The mechanical properties at low temperature indicated that the tensile strength of diphenyl silicone rubber was 4.1Mpa higher than that of methylphenyl silicone rubber at -70 ℃ due to the existence of low-temperature crystallization.

1. Introduction

Silicone rubber is a kind of organic-inorganic hybrid polymer material with Si-O bond as the main chain. Due to the large bond energy (422.5kj /mol), large bond Angle, large orientation degree of freedom and good flexibility, Si-O-Si has super high and low temperature resistance and is widely used in aviation, aerospace, machinery and electronics industries [1]. Commercial methylvinyl silicone rubber generally can work at -55~180℃ for a long time. The phenyl silicone rubber obtained a small amount of Ph2SiO or MePhSiO has better resistance to high and low temperature. Generally, it can work at -70~200℃ for a long time, and can withstand elasticity for several weeks over 200℃ or above 300℃ in an instant. At present, the commercial diphenyl silicone rubber is not easy to widespread application for its complex synthesis process, poor batch stability and the higher price. With the improvement of synthesis technology of methylphenyl cyclosiloxane, methyl phenyl silicone rubber has become a research hotspot due to its simple synthesis process and low cost [2-4]. In this study Methyl Phenyl and diphenyl silicone rubbers were synthesized from octamethyl cyclotetrasiloxane, methylphenyl cyclotetrasiloxane, octaphenyl cyclotetrasiloxane, etc. The high and low temperature properties of silicone rubber were studied by thermogravimetry analysis (TGA), DMTA (dynamic mechanical thermal analysis) and material high and low temperature testing, which provides reference for the development and use of phenyl silicone rubber.

2. Experimental

Materials

D_4(≥99.5%) and tetravinyltetramethylcyclotetrasiloxane(D_4^{Vi}, ≥99.5%), were purchased from Wake Co., Ltd. Methylphenyl cyclosiloxane (≥99.9%) was purchased from Xiantao Greeen Chenical Co., Ltd. Octaphenyl cyclotetrasiloxan (≥99%) was supplied by Dalian Yuanyong Slicone Co., Ltd. Methylvinyl silicone rubber, 110-2VT, was supplied by of China National BlueStar (Group) Co Ltd, Chengrand Research Institute of Chemical Industry Co. Ltd.

Sample Preparation

Synthesis of methyl Phenyl silicone gum

The ring-opening reaction of methylphenyl cyclosiloxane, D4 and D_4^{Vi} were catalyzed by tetramethyl ammonium hydroxide base rubber under nitrogen protection. At the end of the reaction, the catalyst was destroyed, and the small molecules were removed by vacuum decompress. Colorless and transparent methyl phenyl silicone gum (MePhSiO mole fraction 5%, 121-1) was synthesized according this method.

Synthesis of diphenyl silicone gum

The high purity dimethyl diphenyl cyclosiloxane mixture was prepared by catalytic rearrangement of octaphenyl cyclotetrasiloxan and D_4. The mixture cyclosiloxane were catalyzed by tetramethyl ammonium hydroxide base rubber under nitrogen protection with D4 and D_4^{Vi}. Diphenyl silicone gum (Ph_2SiO mole fraction 5%, 120-1) was synthesized according the above method.

Preparation of silicon rubber

According to the literature [1], silicone gum, SiO_2, Fe_2O_3 were compounded and then vulcanized at 160℃ for 15 min under 9.8 MPa of pressure. The vulcanized samples were post cured at 200℃ for 4 h. All the results are taken as the average of 5 samples for each testing.

Measurements

Mechanical tensile tests under various environments were performed on a universal testing machine (Instron3366), at 25°C and around 60% RH. The samples were tested at ambient temperature for 20min, and the stress-strain curve was automatically drawn. The loading speed of the test was 500mm/min. TGA was performed using Q600SDT thermal analysis at a heating rate of 5°C /min in an atmosphere of nitrogen. DMTA were done using IV dynamic mechanical thermal analysis at a heating rate of 2°C /min (frequency, 1 Hz).

3. Results and discussion

Mechanical property. Mechanical properties, hot air aging properties and compression cold resistance coefficient of methyl vinyl silicone rubber (110-2VT), diphenyl silicone rubber (120-1) and methyl phenyl silicone rubber (121-1) are shown in table 1.

Table 1. Properties of silicone rubber

Testing	1	2	3
	110-2VT	120-1	121-1
Tensile strength/MPa	10.6	10.5	10.3
Elongation at break /%	184	113	121
Hardness (shore A)	60	61	63
225℃×72h thermal air aging test			
Rate change of Tensile strength	-20	-12	-14

/%			
Rate change of Elongation at break/%	-36	-27	-31
Rate change of Hardness /%	+3	+2	+2
compression cold resistance coefficient			
-50℃	0.55	0.64	0.71
-70℃	0	0.45	0.61

From table 1, it can be found that, methyl phenyl silicone rubber 121-1 and diphenyl silicone rubber 120-1 have similar tensile strength and slightly lower elongation at break compared with the methyl vinyl silicone rubber 110-2VT. It also showed that a small amount of phenyl structure has a little influence on the tensile strength of silicone rubber, but as for rigid phenyl groups, the flexible order is Me2SiO＞MePhSiO＞Ph2SiO for dimethyl silica link or greater acuity diphenyl methyl phenyl silicone oxygen link link, the higher flexible is corresponding to the bigger elongation performance. The thermal air aging test shows that the phenyl structure improves the thermal resistance of silicone rubber, and the thermal resistance of 120-1 is slightly better than that of 121-1. The compression cold resistance coefficient test shows that the phenyl structure greatly improves the cold resistance of silicone rubber, and the cold resistance of 121-1 is significantly better than that of 120-1.

TGA analysis.

The TG curves of the three silicone rubbers are shown in Fig. 1. It can be seen from Fig. 1 that the initial thermal decomposition temperature curves of the three silicone rubber are almost identical, which indicated that the thermal resistance is not significantly different. With the increase of temperature, the heat resistance 120-1 and 121-1 showed higher heat resistance than 110-2VT, and 120-1 is better than that of 121-1. It also can be seen from Fig. 1 that phenyl structure may prevent main chain crosslinking or degradation [3] caused by oxidative decomposition of the side groups, which improved the thermal stability of silicone rubber. The thermal stability of Ph2SiO is better than that of MePhSiO.

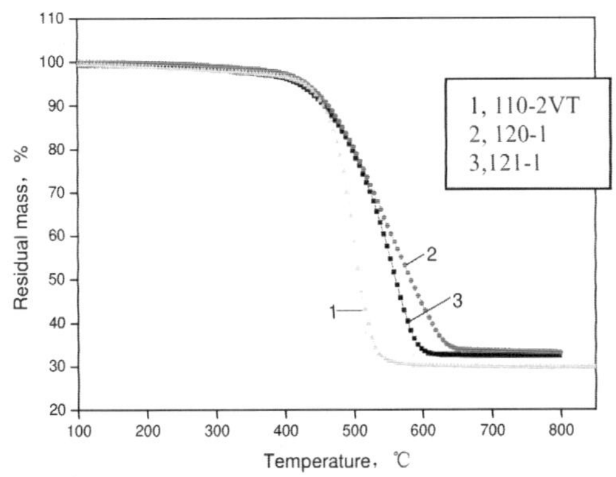

Figure. 1 Thermogravimetric curves for the thermal degradation of silicone rubber samples

Table 2 compares the mechanical properties of 110-2VT, 120-1 and 121-1 at high temperatures. From table 2 it can be seen that with the increase of temperature, the tensile strength and elongation at break of the three silicone rubber samples decreased significantly. This may be due to the fact that the molecular chains of silicone rubber are very flexible and the interaction between the chains is weak. When stretched at high temperature, under the influence of heat effect and volume expansion effect, the motion of polysiloxane molecular chain segment is accelerated, which weakens the force between SiO_2 and polysiloxane molecular chain, resulting in lower tensile strength and elongation at break. And a few phenyl groups have little effect on the mechanical properties of silicone rubber at high temperature.

Table 2. Mechanical properties of samples at high temperature

samples	Temperature /℃	Tensile strength /MPa	Rate change of Tensile strength /%	Elongation at break /%	Rate change of Elongation at break /%
110-2VT	30	10.6	-	184	-
	60	9.6	-9.4	159	-13.5
	90	8.7	-17.9	130	-29.3
	120	6.3	-40.1	99	-46.2
	150	4.4	-58.4	94	-48.9
	200	3.3	-68.9	72	-60.8
120-1	30	10.5	-	113	-
	60	8.3	-20.9	100	-11.5
	90	7.2	-31.4	90	-20.4
	120	5.6	-46.7	72	-36.3
	150	4.7	-55.2	58	-48.7
	200	3.1	-70.5	51	-54.8
121-1	30	10.3	-	121	-
	60	8.0	-22.3	106	-12.4
	90	6.7	-35.0	93	-23.1
	120	5.2	-49.5	73	-39.7
	150	4.8	-53.4	64	-47.1
	200	2.9	-71.8	54	-55.4

Low temperature resistance. DMTA analysis of low temperature resistance of silicone rubber samples is shown in Fig. 2. Elastic modulus (E ') is an important index of rubber elasticity. It can be seen from figure 2, 110-2VT begins to produce crystallization near about -35℃. As the temperature decreases, the crystallinity increases and the elastic modulus increases. 110-2VT belongs to crystalline rubber 120-1 destroyed the regularity of methyl vinyl silicone rubber, resulting in a decrease in crystallization temperature. From -50 degrees, a small amount of tension-oriented crystallization produced, and the elastic modulus increased first and then decreased. 121-1 does not have any crystalline transition,and it performs better at low temperature.

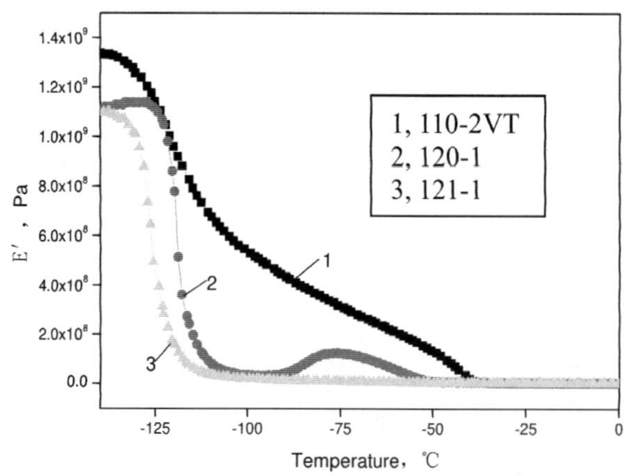

Figure. 2 DMTA analysis of low temperature resistance of silicone rubber samples

The polymer chain structure is the main factor to determine the basic properties of the polymer, while the polymer aggregation structure is the main factor to determine the properties of the polymer. The aggregation form of molecular chain in silicone rubber has great influence on the properties of silicone rubber, and the main factor that affects the aggregation form of silicone rubber is temperature. Figure 3 shows the stress-strain curves of 110-2VT, 120-1 and 121-1 at low temperature, respectively.

It can be seen from the tensile stress-strain diagram of the three silicone rubber samples that the tensile strength and elongation at break of phenyl silicone rubber increase gradually with the decrease of temperature in the range of -70℃ to 30℃. When stretched at low temperature, with the decrease of temperature, the thermal motion energy of polysiloxane molecular chain decreases, and the activity of the motion unit decreases, which is the crystallization effect and volume contraction effect of polysiloxane molecular chain at low temperature. Due to the phenyl structure, the regularity of molecular chain is destroyed and the crystallinity of silicone rubber (120-1 and 121-1) is greatly reduced in the range of -70℃~30 ℃. In the range of -30℃~-70℃, the slope of the curve increases obviously, and low temperature crystallization is the main reason leading to the increase of 110-2VT. Affected by the combined influence of crystallization and volume shrinkage of silicone rubber, the mechanical properties of 110-VT increased first and then decreased. The mechanical properties of 120-1 and 121-1 showed obvious enhancement (Table 3). 120-1 is greatly affected by low-temperature crystallization effect, and it shows higher tensile strength (20.7 MPa) at -70℃, which is 4.1MPa higher than 121-1 (16.6 MPa).

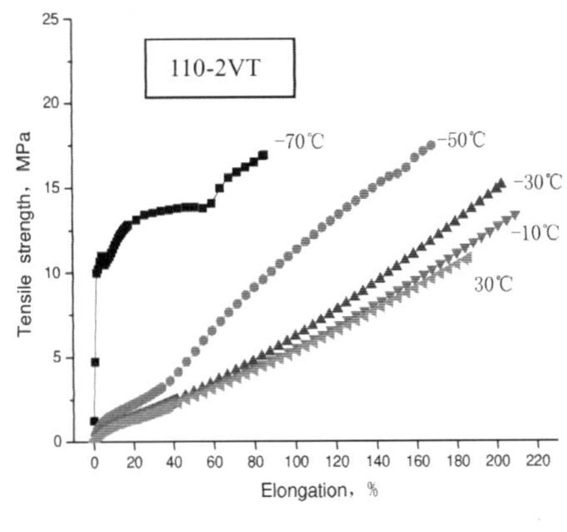

110-2VT

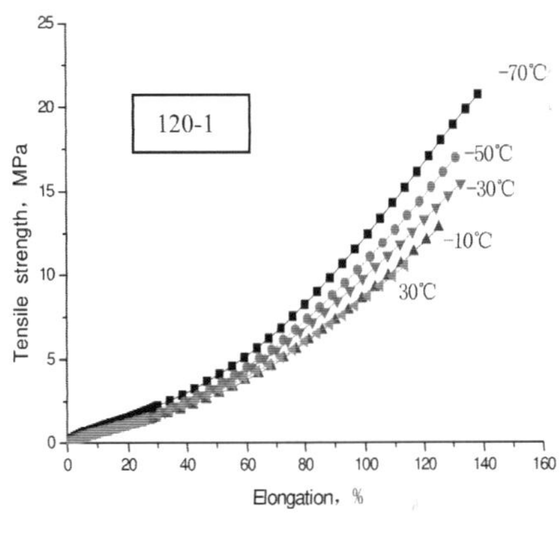

120-1

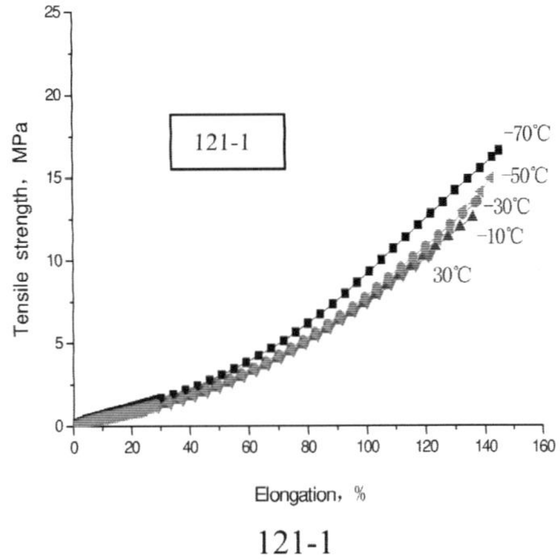

121-1

Figure. 3 The stress-strain curves of 110-2VT, 120-1 and 121-1 at low temperature

Table 3. Mechanical properties of samples at low temperature

samples	Temperature /℃	Tensile strength /MPa	Rate change of Tensile strength /%	Elongation at break /%	Rate change of Elongation at break /%
110-2VT	30	10.6	-	184	-
	-10	13.4	+26.4	209	+13.6
	-30	15.1	+42.5	202	+9.8
	-50	17.4	+64.2	167	-9.2
	-70	16.9	+59.4	84	-54.3
120-1	30	10.5	-	113	-
	-10	12.8	+21.9	125	+10.6
	-30	15.4	+46.7	132	+16.8
	-50	16.9	+61.0	130	+15.0
	-70	20.7	+97.1	138	+22.1
121-1	30	10.3	-	121	-
	-10	12.5	+21.4	135	+11.6
	-30	13.5	+31.1	137	+13.2
	-50	14.2	+37.9	141	+16.5
	-70	16.6	+61.2	145	+19.8

4. Conclusion

The reliability of steel structural roof is computed under big data loads. According to the safety rules of the building and serviceability objectives, the Monte Carlo method is used to consider the uncertainty of roof load and the big data loads on the roof. The reliability of the steel structural roof under the big data loads was obtained. However, it should be noted that this reliability assessment is usually less than 25-45%. Although the reliability is similar, even if the roof design has the same load exceeding 50 years of the load, the substantial difference in probability shape will result in the distribution of the maximum snow load per year. Producing different reliability results, these are important for roof snow loads.

In additional, the application of this analysis method should be fully consider the difference climate and geological distribution. The annual maximum snow load has a great impact on the reliability analysis. At the same time, changing the physical characteristics of the roof has little effect on the reliability index. Secondly, there is a strong correlation between the reliability and maintainability of steel roofs. Finally, for the steel structural roof, the design of the deflection limit results in a larger cross-sectional dimension, resulting in considerable improvement in safety, maintainability and reliability.

References

[1] S.Feng, J. Zhang,M.Li, et al., Organosilicon polymer and application. Chemical Industry Press, Beijing, 2004, pp.101-105.
[2] H.Zhang, D.Cao, J.Wang, et al., High performance rubber sealing material . Beijing, Chemical Industry Press, Beijing, 2007, pp.79.
[3] K.A. Andrianov, G.L. SlonimskiĬ, A.A. Zhdanov, et al., Some physical poperties of polyorganosiloxanes. Ⅰ. linear polyorganosiloxanes. J. Polym. Sci. : Part A. 10 (1972) 1.
[4] K.A. Andrianov, G.L. SlonimskiĬ , A.A. Zhdanov, et al., Some physical properties of polyorganosiloxanes. Ⅱ. nonlinear and irregular polyelementorganosiloxanes. J. Polym. Sci.: Part A. 10(1972) 23.

Acknowledgments

This research was financially supported by Key R & D project of Shandong Province (grant No.2019JZZY010350)

Effect of cryogenic treatment of cemented tungsten carbide tools for die application

C Yooliengpun[1], N Bandasak[1], K Tuchinda[2, 3] and O Diewwanit[1]

[1] Department of Tool and Materials Engineering, King Mongkut's University of Technology Thonburi, Bangkok 10140, Thailand
[2] The Sirindhorn International Thai-German Graduate School of Engineering, King Mongkut's University of Technology North Bangkok, Bangkok 10800, Thailand
[3] Material Manufacturing and Surface Engineering Research Center, Science and Technology Research Institute, King Mongkut's University of Technology North Bangkok, Bangkok 10800, Thailand

onnjira.tha@kmutt.ac.th

Abstract. This study investigated effects of cryogenic treatment and post tempering on microstructure parameters and mechanical properties of cemented tungsten carbide for die applications. Carbide grain size was increased after the cryogenic treatment resulted in an increased carbide contiguity. After tempering process, tungsten carbide grain size was reduced and more uniform in size distribution with less in WC-Co debonding. The fracture toughness of cryogenically treated carbide which evaluated from microstructure parameter was slightly increased from 20 to 21.1 MPa·m$^{1/2}$. After tempering, the fracture toughness was almost unchanged compared to that of untreated sample and the number of tempering cycles did not affect the fracture toughness. However, hardness and scratch resistance were improved 10% and 60%, respectively, by cryogenic treatment. Increasing the number of post tempering cycles after applying the cryogenic treatment result in reducing hardness but did not affected the scratch resistance of the cemented tungsten carbide. In addition, results of ball on disc tribological testing indicated that wear resistance of the cemented tungsten carbide could be improved by cryogenic treatment due to the higher hardness and scratch resistance as expected. The wear volume was increased with the number of tempering cycles. Therefore, sliding wear resistance and scratch resistance applications in which material surface failure behaviour is mainly controlled by surface hardness, post tempering process might not be required.

1. Introduction

Cemented tungsten carbides or hard materials are wildly used in hard and high-quality wear resistance applications. Beside the use for cutting tools, they are also used as moderate-impact and high-impact dies such as wire drawing, bunching, stranding, guiding, and split die applications [1]. This is dues to hardness, strength and fracture toughness of the materials. It is well known that the mechanical properties of the cemented tungsten carbide mainly depend on an average grain size, cobalt content and processing to achieve microstructure [2, 3]. Typical microstructure of the cemented tungsten carbide composed of WC, TiC, TaC embed in the cobalt binder matrix. In addition, a ternary compound of tungsten, cobalt and carbon (η phase) can exist during the cooling state of sintering [2]. Previous studies indicate that using cryogenic treatment was mainly on improving wear resistance whereas life of the tungsten carbide cutting tools was well documented [4-7]. The cryogenic treatment is generally performed in two methods, shallow and deep practice. The shallow cryogenic treatment is executed at temperatures in the range -80°C whereas deep cryogenic treatment is carried out at temperatures between -140 and -196 °C [8, 9]. The cryogenic treatment process involves slowly cooling materials to the cryogenic, holding it for a certain period of time and then gradually warming back to room temperature. Controlled cryogenic treatment increases values of fineness, uniforms distribution and densification of cobalt binder. It does not affect crystal structure of tungsten carbide (WC) particles and consequently, it improves hardness and bending strength of the cemented carbide [10]. Overall the treatment makes the carbides more firm and better wear resistance. Dhande et al. [11] studied the effect of soaking periods in cryogenic treatment of tungsten carbide insert. They found that 8 hrs. soaking period at -185 °C brought about the improvement in the wear characteristics that obtained by pin on disc wear test. Ozbek et al. [12] showed that the carbide inserts which cryogenically treated at −145 °C for 24 hrs. exhibited the best wear resistance in turning stainless steel. Formation of W_2C and Co_3W_3C (secondary carbides), alongside with cobalt binder create a compressive residual stress and tough matrix [5]. These can improve wear resistance of the carbide cutting tools. Ozbek et al. [13] stated that the treated inserts in dry turning stainless steel showed a better performance than the untreated ones of up to 34% and 53% in terms of frank wear and crater wear, respectively. Arun et al. [4] mentioned that deep and shallow cryogenically treated drill exhibit equivalent performance in drilling small series of holes. As a result, shallow cryogenically treated drill would be more economical working. In addition, effect of tempering followed cryogenic treatment has been studied together with the tungsten carbide inserts [14-16]. Deshpande et al. [14] and Kalsi et al. [15] found that tempering had a significant influence on the phases presented in WC-Co inserts without changing the surface morphology and subsequently influenced their machining performance. The cryogenic treatment improved the microhardness of the tungsten carbide inserts but the microhardness was decreased with an increase in the number of tempering cycles after the treatment [15]. In addition, the cryogenically treated insert with double and triple tempering cycles exhibited the highest wear resistance. However, Kalsi et al. [15] indicated that the fourth tempered insert reduced effect of cryogenic treatment by disruption in the cobalt binder. Recently, Weng et al. [17] found that deep cryogenic treatment could improve both fracture toughness and wear resistance of WC-20Co cemented carbide for die applications. However, scientific literature in the area of improvement mechanical properties of the cemented tungsten carbide for die applications is limited. No study is evident about effectiveness of post tempering and number of tempering cycles on microstructure and mechanical properties of the cemented tungsten carbide in case of die applications that emphasize fracture toughness and high wear resistance. The objectives of this investigation are to evaluate the effects of cryogenic treatment and post tempering cycles on microstructure parameter and mechanical properties; hardness, scratch resistance and wear analysis of the cemented tungsten carbide for die applications.

2. Materials and Methods

2.1 Materials

The cemented tungsten carbide (80 wt.%WC-20 wt.%Co) were used as samples of this study and classified as highly impact tool and die applications.

2.2 Cryogenic treatment and post tempering

The samples were cryogenically treated with step cooling at -140 ℃ and held at this temperature for 24 hrs. After the cryogenic treatment (CT), the samples were gradually brought to room temperature, then submitted to different tempering treatments; single (1T), double (2T) and triple (3T) tempering at 200 ℃ for 2 hrs. to relieve stresses induced in cryogenic treatment.

2.3 Microstructure Analysis

Microstructure analysis of the WC-Co cemented carbide was carried out under Scanning Electron Microscope (JEOL, Japan). The samples were polished by diamond paste (1 μm) to reach mirror finish and etched with KOH reagent. To study effects of the cryogenic treatment, microstructure parameters were evaluated. This included carbide grain size distribution, average carbide size, carbide contiguity (WC/WC interface), and binder mean free path for each sample. All of these were evaluated from the 9 SEM photomicrographs. Fracture toughness, K_{1C} of the WC-Co cemented carbide was calculated by these microstructure parameters.

2.4 Hardness Test

Measurements of microhardness was performed under a load of 300 g for 15 s by using a micro-Vickers hardness tester (FUGITEC, Japan). The microhardness was measured at 13 points for each sample and mean value of these measurements was accepted as microhardness value of the samples.

2.5 Scratch Test

Scratch resistance of the samples was carried out on a scratch tester (CSM Instruments, Switzerland) with diamond indenter and with applied progressive loads from 0.9 N to 100 N. Scratching speed was 10 mm/min and scatch length was 10 mm.

2.6 Wear Test

Sliding wear test was performed using a Ball on disc tribological testing machine (CSM Instruments, Switzerland) at room temperature under non-lubricated condition. The cemented tungsten carbides were prepared as the disc and the ball made of cemented tungsten carbide coated with AlCrTiN. A load of 15 N was imposed that created the maximum 3,050 MPa in contact pressure. Sliding velocity was fixed at 0.1 cm/s, while cycles were changed from 1500 to 2000, 2500, 3000, and 3500. A contact radius of 3 mm was applied. A profilometer was used to map the contour of the worn surfaces. Wear volume was subsequently determined.

3. Results and Discussion

3.1 Microstructural analysis and fracture toughness evaluation

Figure 1 shows SEM pictures of the untreated, the cryogenically treated and the cryogenically treated and single tempered tungsten carbide samples. Carbide grain size is one factor that affects the mechanical properties and wear resistance. To study the effect of cryogenic treatment, sizes of carbide grains were randomly measured by using the line intercepts method from 9 images, 15 grains per image, in order to compare the average grain size of the untreated with cryogenically treated samples. The average of carbide grain sizes of the samples and standard deviation with various conditions were shown in Figure 2. Upon measuring sizes of the carbide grains, it was found that the average grain size of the untreated cemented tungsten carbide was 4.32 μm and the cryogenically treated tungsten carbide was 4.5 μm, respectively. Cryogenic treatment led to an increase of 4% in the grain size and greater standard deviation of the treated tungsten carbide with respect to that of the untreated one. As mention by Thornton [18], the increasing of tungsten carbide grains reduces micro porosity and results in a more contiguous microstructure. The larger grain size caused an increase in carbide contiguity. This agree well with that found in case of the tungsten carbide insert tool. However, the grain size was slightly reduced to 4.30 μm after the tempering and the standard deviation was also reduced. This can be interpreted that the carbide of the sample is more uniform in size after the tempering. Increasing in the number of tempering cycles does not affect the carbide grain size but it becomes more uniform in size distribution. Microstructure of the untreated and the cryogenically treated tungsten carbide was shown in Figure 1(a) and 1(b), respectively. It can be observed that WC debonding and fragment of WC grains became less in cryogenically treated sample. In case of the single tempered sample as illustrated in Figure 1(c), the WC debonding is hardly found but small torn off WC are existed.

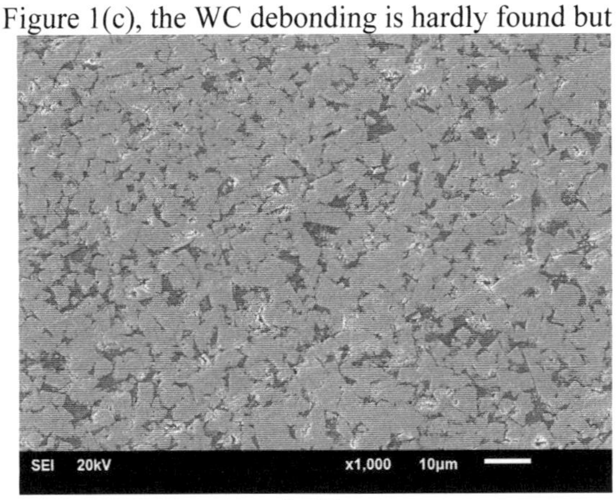

(a-1)

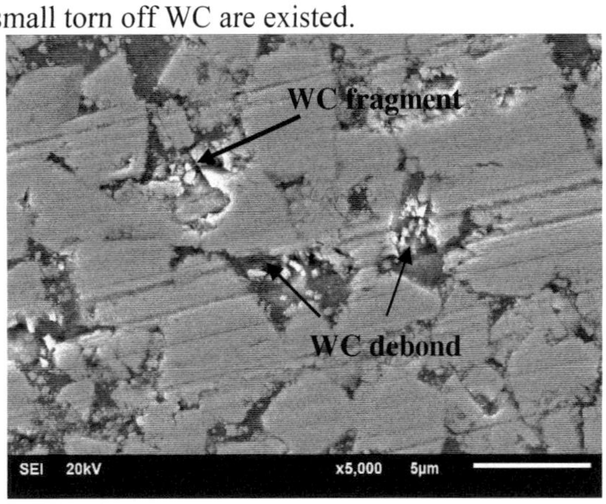

(a-2)

IOP Conf. Series: Materials Science and Engineering 1048 (2021) 012002 doi:10.1088/1757-899X/1048/1/012002

(a)

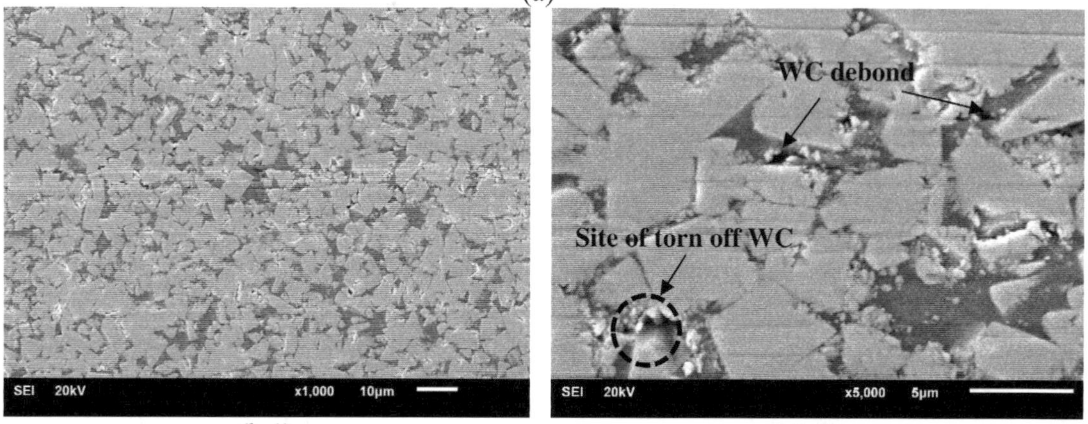

(b-1) (b-2)

(b)

(c-1) (c-2)

(c)

Figure 1. Microstructure of cemented tungsten carbide (a) untreated (b) cryogenically treated (c) cryogenically treated with single tempered.

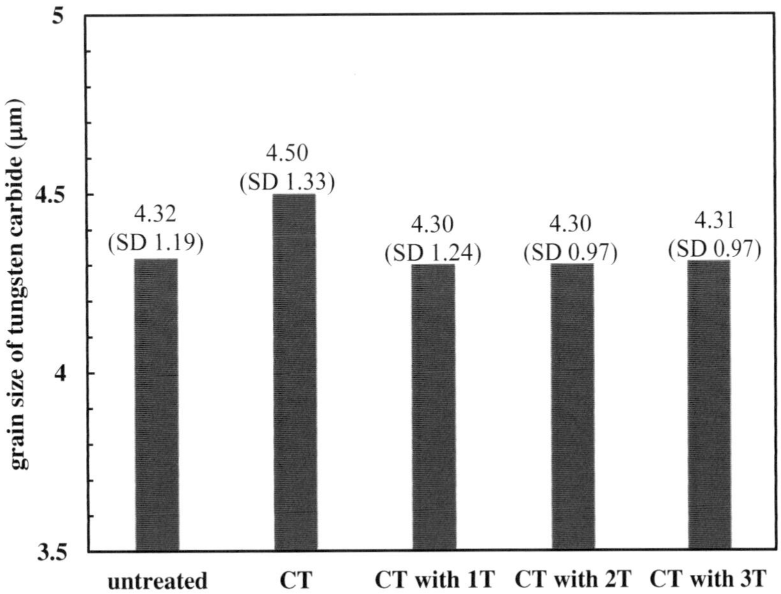

Figure 2. Effect of cryogenic treatment and post tempering on average grain size of tungsten carbide.

Microstructure parameters which are the carbide contiguity and binder mean free path of the cemented tungsten carbide were evaluated and also shown in Table 1. The carbide contiguity indicated the degree of contacts between carbide grains. It was found that cryogenic treatment increased the carbide contiguity from 0.82 to 0.87. These might be a result of the cobalt binder contraction related to the cryogenic treatment. Consequently, the carbide particles become closer. In case of the single tempered sample, carbide contiguity was unchanged, comparing with the cryogenically treated one. The cobalt binder was slightly expanded during the double and triple tempering, and the carbide contiguity was reduced to 0.81.

Table 1. Microstructural parameter and K_{IC} of cemented tungsten carbide.

	Carbide average size (μm)	Carbide contiguity	Binder mean free path (μm)	K_{IC} ($MPa \cdot m^{1/2}$)
Untreated	4.32 ± 1.19	0.82	1.97	20.7
CT	4.50 ± 1.33	0.87	2.11	21.1
CT with 1T	4.30 ± 1.24	0.87	2.01	20.7
CT with 2T	4.30 ± 0.97	0.81	1.94	20.6
CT with 3T	4.31 ± 0.97	0.81	1.95	20.7

Fracture toughness, K_{IC} of the cemented tungsten carbide, is known to increase as binder phase volume fraction and average carbide grain size. The binder phase mean free path is increased, while decreased carbide contiguity causes a decrease in the fracture toughness. K_{IC} of the cemented carbide can be evaluated from microstructure parameters, as suggested by Gopal [2] and Alex et al. [19] :

$$K_{IC} = 3.907 + 0.325V_B\% + 2.389\bar{d}_C - 0.878\lambda + 2.065C$$

$V_B\%$, $\bar{d}_C$, λ and C are vol% binder phase, average carbide grain size, binder phase mean free path and carbide contiguity, respectively. The fracture toughness, K_{IC} of the samples as evaluated from the microstructural parameters are also shown in Table 1. The K_{IC} of the cryogenically treated tungsten carbide was higher than that of the untreated tungsten carbide. These might be due to the increased grain size and carbide contiguity which are the microstructure parameter which promoted the K_{IC}. After the single tempering, the fracture toughness was almost unchanged and the number of tempering cycles did not affect the fracture toughness but the microstructure became more uniform.

3.2 Hardness

Hardness of the untreated and cryogenically treated cemented tungsten carbide are shown in Table 2. The data indicates that, average hardness of the untreated sample is 989.02 HV and the cryogenically treated sample is 1079.93 HV with high value of standard deviation. Since, the post tempering process affected hardness of the cryogenically treated samples, hardness of the cryogenically treated sample was significantly reduced after the single tempering. Standard deviation of the hardness was also reduced after tempering process. This indicated that: the tempered samples are more uniform in hardness; increasing the number of tempering cycles bring about the slightly decreasing in hardness; and tempering process reduces residual stress without changing the microstructure.

Table 2. Hardness of the cryogenically treated and tempered cemented tungsten carbide.

	Hardness (HV)	**Standard deviation**
Untreated	989.02	23.04
CT	1074.93	32.66
CT with 1T	962.47	14.61
CT with 2T	951.81	15.51
CT with 3T	947.86	7.89

3.3 Scratch Resistance

Studying effects of the cryogenic treatment, fracture toughness and hardness on the scratch resistance of the cemented tungsten carbide were evaluated in terms of critical loads for the appearance of first cracks (LC1), dense cracking pattern (LC2) and completed cracks (LC3) [20]. The critical loads under the scratch test are shown in Figure 3. The scratch resistance properties of the cemented tungsten carbide could be improved by the cryogenic treatment and the tempering process. First crack in the cryogenically treated samples could not observed up to the critical load (LC1) of 56.02 N and dense cracking pattern could not be detected up to the maximum load of 66.96 N. The fracture crack occurred at 88.81 N. These evidence higher scratch resistance of the treated tungsten carbide comparing with the untreated cemented tungsten carbide that the first crack was occurred at the critical load (LC1) of 35.42 N. This is because of the reduced WC debonding and fragment of WC grains in cryogenically treated sample as illustrated in Figure 1(b). In case of the cryogenically treated and single tempered tungsten carbide which less WC debonding and fragment, earlier cracks were observed up to the critical load (LC1) about 55 N which closed to the LC1 of the cryogenically treated samples. The number of tempering cycles did not cause much changes in the critical loads for the appearance of the first crack. However, the cryogenically treated samples were completely cracked under the highest load. These might be due to the highest hardness of the cryogenically treated sample. The experimental results were consistent with WC-Co carbide with a high cobalt content. The dominant damage mechanisms are the plastic deformation of the WC grains via slip and the formation of intergranular cracks [21]. Therefore, the scratch resistance of the cemented tungsten carbide could be improved by the cryogenic treatment and tempering cause that the debonding of tungsten carbide decreased.

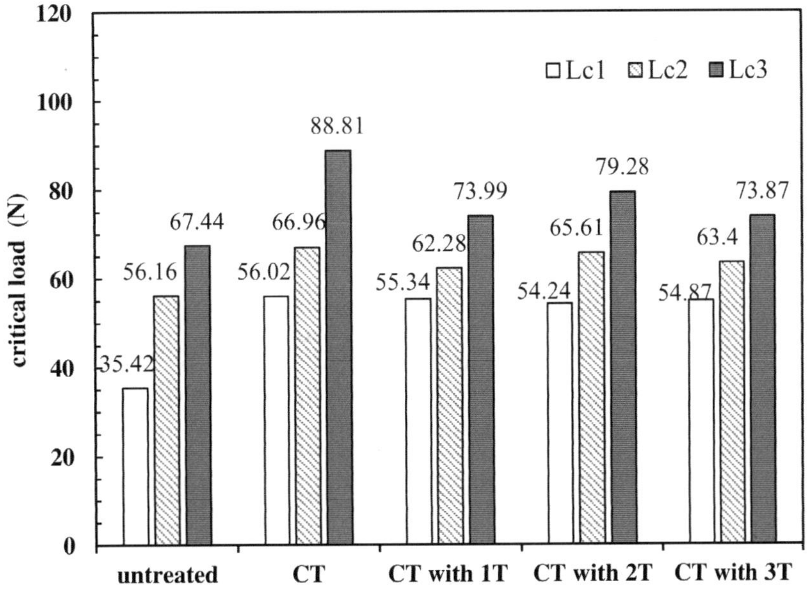

Figure 3. Effect of the cryogenic treatment and post tempering on critical load under scratch resistance test of cemented tungsten carbide.

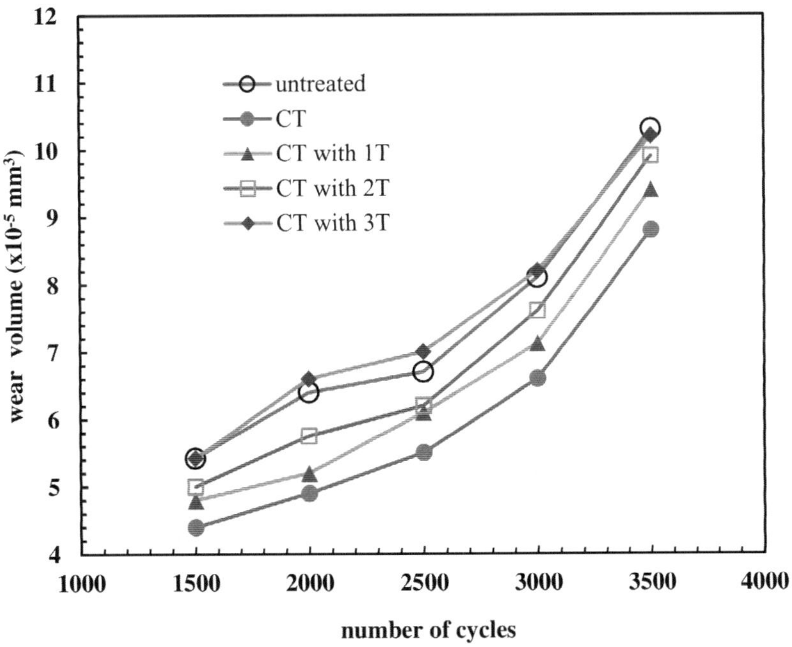

Figure 4. Evolution of wear volume with number of cycles of cemented tungsten carbide.

3.4 Wear resistance

An illustration of the effect of the cryogenic treatment and tempering on wear volume of the cemented tungsten carbide is shown in Figure 4. The wear volume was increased as the number of the test cycles. The lowest wear volume was obtained in the case of cryogenically treated cemented tungsten carbide with the high hardness of 1074.93 HV and also displayed the highest fracture toughness 21.1 MPa·m$^{1/2}$.

This is due to the densification of the cobalt binder, which induces a compressive stress on the WC grains, resulting in an increase in wear resistance. The wear volume of the cryogenically treated with single tempered tungsten carbide was slightly greater than the cryogenically treated one. However, increasing the number of tempering cycles brought about the increasing wear volume. The cryogenically treated with triple tempered cemented tungsten carbide showed the highest wear volume which closed to the wear volume of the untreated cemented tungsten carbide. These might be a result of the triple tempering after the cryogenic treatment produced disruptions in the cobalt binder as found in case of carbide insert tool [9]. Increasing fracture toughness on the expense of hardness did not affect wear resistance.

4. Conclusions

The study investigated effects of cryogenic treatment and post tempering cycles on microstructure parameters and mechanical properties of cemented tungsten carbide for die applications. It was found that cryogenic treatment and single tempering process promoted a densification of cobalt binder. Single tempering after cryogenic treatment promoted the uniform grain size and carbide contiguity of the cemented tungsten carbide. The hardness and fracture toughness, K_{IC} of the cryogenically treated tungsten carbide, were increased with increasing mean free path in the cobalt binder phase and increasing carbide contiguity. Hardness of the cemented tungsten carbide decreased with an increase in the number of tempering cycles after applying the cryogenic treatment. The cryogenically treated tungsten carbide exhibited the highest scratch resistance due to higher surface hardness. Cryogenic treatment improved wear resistance of the cemented tungsten carbide, however, wear volume increased with the number of tempering cycles. The cryogenically treated with triple tempered cemented tungsten carbide showed the highest wear volume which closed to the wear volume of the untreated cemented tungsten carbide. The investigation indicated that hardness is still the main factor affecting sliding wear resistance under the current testing conditions and surface scratch resistance. Tempering after the cryogenic treatment does not increase fracture toughness but result in lower hardness. However, post tempering could promote a uniform in microstructure and hardness could be advantaged for die application with impact wear and high local stress area. For sliding wear resistance and scratch resistance applications in which material surface failure behavior is mainly controlled by surface hardness, post tempering process might not be required.

References

[1] Davis J R 1995 *ASM Specialty Handbook: Tool Materials*. (Ohio, U.S.A: ASM International).
[2] Gopal S U 1998 Microstructural aspects of cemented carbide, *Cemented tungsten carbides: production, properties and testing* (New Jersey, U.S.A.: Noyes Publication) 166-92.
[3] Ndlovu S 2009 The wear properties of tungsten carbide-cobalt hardmetals from the Nanoscale up to the macroscopic scale (Doctoral thesis, University of Erlangen-Nuremberg, Erlangen, Germany).
[4] Arun M, Arunkumar N, Vijayaraj R and Ramesh B 2018 Investigation on the performance of deep and shallow cryogenic treated tungsten carbide drills in austenitic stainless steel. *Measurement* **125** 687-93.
[5] Gill S S, Singh J, Singh H and Singh R 2012 Metallurgical and mechanical characteristics of cryogenically treated tungsten carbide (WC–Co). *International of Advanced Manufacturing Technology* **58** 119-31.
[6] Sreerama R, Sornakumar T, Venkatarama R and Venkatram M 2009, Machinability of C45 steel with deep cryogenic treated tungsten carbide cutting tool inserts. *International Journal of Refactory Metals and Hard Materials* **29** 181-85.
[7] Yong A. Y. L, Seah K H W, and Rahman M 2006 Performance evaluation of cryogenically treated tungsten carbide tools in turning. *International Journal of Machine Tools and Manufacture* **46** 2051-56.
[8] Baldissera P and Delprete C 2008 Deep cryogenic treatment: A bibliographic review. *The Open Mechanical Engineering Journal* **2** 1-11.
[9] Kalsi N S, Sehgal R and Sharma V S 2014 Effect of tempering after cryogenic treatment of tungsten carbide-cobalt bounded inserts. *Bullatin of Materials Science* **37(2)** 327-35.

[10] Razavykia A, Delprete C and Baldissera P 2019 Correlation between microstructural alteration, mechanical properties and manufacturability after cryogenic treatment: A review. *Materials* **12(3302)**.
[11] Dhande S T, Kane V A, Dhobe M M, and Gogte C L 2018 Influence of soaking periods in Cryogenic treatment of tungsten carbide. *Procedia Manufacturing* **20** 318-28.
[12] Ozbek N, Cicek A, Gulesin M and Ozbek O 2014 Investigation of the effects of cryogenic treatment applied at different holding times to cemented carbide inserts on tool wear. *International Journal of Machine Tools and Manufacture* **12** 34-43.
[13] Ozbek N, Cicek A, Gulesin M and Ozbek O 2016 Effect of cutting conditions on wear performance of cryogenically treated tungsten carbide inserts in dry turning of stainless steel. *Tribology International* **94** 223-33.
[14] Deshpande R G and Venugopal K A 2018 Machining with cryogenically treated carbide cutting tool inserts. *Materials Today: Proceedings* **5(1)** 1872-78.
[15] Kalsi N S, Sehgal R and Sharma V S 2010 Cryogenic treatment of tool materials: A review. *Materials and Manufacturing Processes* **25** 1077-1100.
[16] Thakur D, Ramamoorthy B and Vijayaraghavan L 2008, Influence of different post treatments on tungsten carbide–cobalt inserts. *Materials Letters* **62** 4403-06.
[17] Weng Z, Gu K, Liu X, Cai H and Wang J 2019 Effect of deep cryogenic treatment on the fracture toughness and wear resistance of WC-Co cemented carbides. *International Journal of Refactory Metals and Hard Materials* **85** 1-6
[18] Thornton R W 2014 investigating the effects of cryogenic processing on the wear performance and microstructure of engineering materials (Doctoral thesis, The University of Sheffield, U.K.).
[19] Alex V, Shatov S S, Ponomaver and Firstov S A 2014 *Comprehensive Hard materials*, ed Mari D et al. (U.S.A: Elsevier) 301-44.
[20] Paiva J M, Fox-Rabinovich G, Junior E L, Stolf P, Ahmed Y S, Martins M M, Bork C and Veldhuis S 2018 Tribological and wear performance of nanocomposite PVD hard coatings deposited on aluminum die casting tool. *Materials* **11(358)** 1-15.
[21] Östberg G, Buss K, Christensen M, Norgren S, Andrén H, Mari D, Wahnström G, Reineck I 2006 Mechanisms of plastic deformation of WC–Co and Ti(C, N)–WC–Co. *International Journal of Refactory Metals and Hard Materials* **24** 135-44.

Determination of damping ratio using Monte-Carlo method

Q Peng[1], Y Ding[2], Y Huang[3] and Z Zhu[4]

[1]Master candidate, Hohai University, College of Civil and Transportation Engineering, 210098 Nanjing, China
[2]Ph.D candidate, Université de Lorraine, Laboratoire d'étude des microstructures et de mécanique des matériaux, 57070 Metz, France
[3]Master candidate, Université Paris 1 Panthéon-Sorbonne, 75005 Paris, France
[4]Ph.D candidate, Ecole des Ponts ParisTech, Laboratoire Navier/CERMES, France

E-mail: zhehao.zhu@enpc.fr

Abstract. This paper highlights an innovant method for calculating the damping ratio using Monte-Carlo method. Compared with other approaches reported in the literature, the clear benefit of the method is the ability to directly determinate the damping ratio upon irregular stress-stain curves through Python. By using Monte-Carlo method, it is found that (i) the computed damping ratios are consistent with the undrained cyclic triaxial result on a dense reference sand specimen; (ii) while total random samplings number for each hysteresis loop greater than 5000 times, the method could provide a satisfactory analysis.

1. Introduction

In the field of materials science, damping ratio is an important indicator directly reflecting the capacity of energy dissipation (e.g., concrete [1], steel [2]). In soil mechanics, when subjected to cyclic shearing in undrained condition, sand could undergo a transient loss of shear resistance owing to the rapid build-up of excess pore water pressure. During this process, sand flows almost like a liquid [3], which could be directly identified by the apparent rise of damping ratio.

The damping ratio [4] is defined as $D = \frac{1}{4\pi} \times \frac{\Delta W}{W}$, where ΔW and W stand for energy dissipated and peak energy in each single cycle, respectively. In practice, the damping ratio (D) is commonly determined in stress-strain curves (see Figure. **1**), and its value is proportional to area ratio between the dissipated energy (blue hysteresis loop) and the peak energy (red triangle). In this sense, a traditional method is to sequentially calculate D one after another by assuming that hysteresis loop could be possibly viewed as an ellipse. Unfortunately, this method is extremely time-consuming since the cycle number N_{cyc} is usually huge. In order to solve this problem, Wei [5] deduced that D could be roughly calculated by a simple formula: $D = \frac{1}{2} \times \frac{\sigma_{do}}{\sigma_m}$, where σ_{d0}, σ_m stand for intersect stress value with y-axis and maximum stress in one cycle, respectively. Later, Chen [6] further pointed out that the polygon had a better fitting performance for hysteresis loops, as compared with ellipse. However, numerous experimental studies [7] indicate that the hysteresis loop is definitely irregular, especially in large deformation. The representation of hysteresis loop either by ellipse or polygon is only a rough hypothesis and the precise determination of D remains an open question. Thus, the emphasis of this paper is to present an innovant method, which is able to directly determinate the damping ratio on any irregular stress-stain curves by employing Monte-Carlo method.

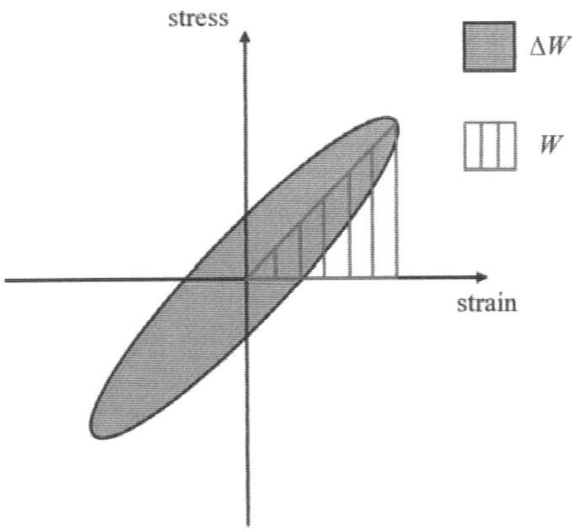

Figure. 1 Relationship between hysteresis loop and damping ratio (after [4])

2. Monte-Carlo method

Originated in the 1940s, Monte-Carlo method (also referred to as "Monte-Carlo experiments") relies on a great amount of repeated random samplings for the purpose of obtaining a probability as the solution to a given event that might be deterministic. Nowadays, with the rapid development of calculation power, the mimicking of random samplings could be quite easily realized in many commercially available platforms. In particular, Python has been widely proved to have greater advantages in this field since it is totally Open-Source and object-orienting. Thereafter, Monte-Carlo method based on Python has gradually become one of the standard tools in physical or mathematical problems when it is extremely difficult or seemingly at least impossible to apply other approaches.

In image processing, this method is commonly employed in calculating the surface of irregular graphics. Thus, the area of irregular hysteresis loops (S_{loop}) linking to damping ratio of sand response is considered as a reference instance in the present work, and the main processes could be divided into the following steps (Figure. **2**):

1. Split the target loop from the complete stress-stain curve with a constant external frame. The boundary values of external frame are respectively x_{min}, x_{max} and y_{min}, y_{max} for x- and y- axes, implying that the surface of external frame S_{tot} is equal to $(y_{max}-y_{min})\times(x_{max}-x_{min})$. During this process, it should be noticed that the after-splitting single loop needs to be strictly enclosed. In other words, the presence of intersection point (represented by solid green point in Figure. **2**) is imperative.
2. Generate random points (x_{rand}, y_{rand}) with a total samplings number T_{ot} in the coupled range of $[x_{min}, x_{max}]$ and $[y_{min}, y_{max}]$.
3. For each random point, interpolate (e.g., linear manner) the upper and lower boundaries of the hysteresis loop in terms of the x-coordinate of random point (x_{rand}) to find the corresponding $y_{int\text{-}min}$ and $y_{int\text{-}max}$.
4. Estimate the relative position of random point to hysteresis loop. If $y_{rand}<y_{int\text{-}max}$ and $y_{rand}>y_{int\text{-}min}$ could be simultaneously satisfied, it could be logically deduced that this random point is an

inside one, illustrated by the hollow red point in Figure. **2**. Note that, in order to render all judgements regarding the relative position achievable, the hysteresis loop needs to be geometrically enclosed, as previously mentioned in the 1st step.

5. Repeat the 3rd and 4th steps until all generated random points are treated.
6. Count the number of inside random points T_{ins}. Then, the probability of these points is equal to T_{ins}/T_{ot} and, at same time, equal to S_{loop}/S_{tot}. Accordingly, the desired hysteresis loop surface (S_{loop}) could then be calculated: $(S_{tot}\times T_{ins})/T_{ot}$.

It is worth noting that although the probability is dependent on the used external frame (e.g., size and shape), the final result (S_{loop}) is still irrespective of the frame. Besides, the more random points there are, the more accurate the final result becomes since one might expect that the probability tends to gradually approach the true value with the increasing number in random points.

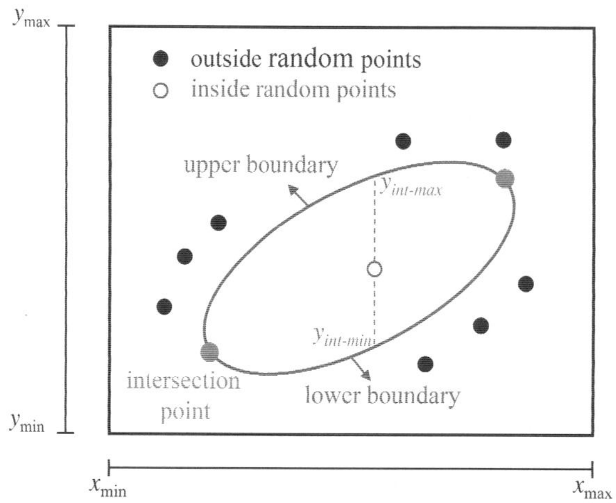

Figure. 2 Illustration of Monte-Carlo method on the calculation of hysteresis loops from undrained cyclic triaxial tests

3. Computation results

The stress-stain curve (q/ε_a: deviator stress/axial strain) of undrained cyclic triaxial tests on a clean dense HN31 sand specimen (density index of sand matrix I_{Dmat}=0.70, cyclic stress ratio T_{cc}=0.22 and effective confining pressure σ'_c=200 kPa) is depicted in Figure. **3**. Note that detailed index properties of the used HN31 sand and experimental program are reported by Zhu [8] and Jradi [9], respectively. Three distinct behaviours can be distinguished as follows:

(a) At the very beginning of test (e.g. $|\varepsilon_a|\leq 0.5\%$), the hysteresis loops are almost linear and coincide with each other. This suggests that in this minor range of ε_a, the sand response could be viewed as "pseudo-elastic" without significant energy dissipation.

(b) With the further development of ε_a, the plasticity of sand becomes much more important, which can be embodied by the increasing discrepancy in the subsequent loops.

(c) In the extremely large range of ε_a (e.g., $|\varepsilon_a|\geq 5\%$), the hysteresis loops become quasi-"Z" shape, exhibiting a strong dissipation capacity. This is caused by the triggering of liquefaction in which the sand specimen behaves almost like a liquid due to the rapid build-up of excess pore water pressure (see the subplot of Figure. **3**).

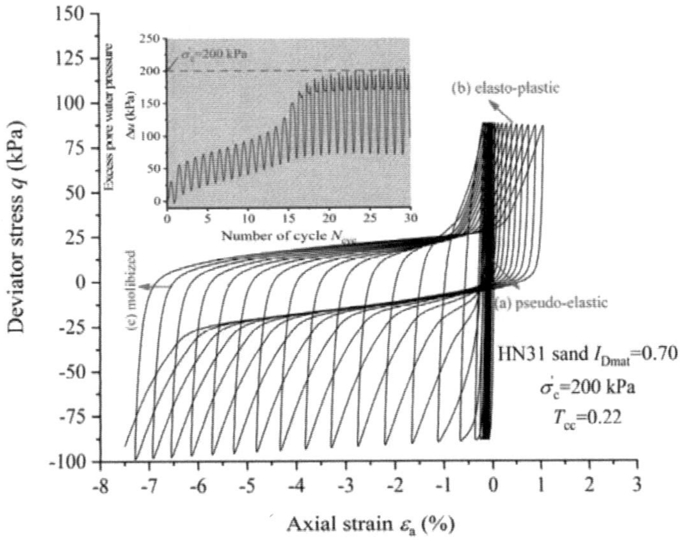

Figure. 3 Three distinct responses of dense HN31 (I_{Dmat}=0.70) sand in undrained cyclic triaxial loading of σ'_c=200 kPa, T_{cc}=0.22: (a) pseudo-elastic; (b) elasto-plastic and (c) mobilized

Due to the constraints of limited space, only several treated hysteresis loops are schematized in Figure. **4** where the damping ratio D (determined by Monte-Carlo method of total random samplings number T_{ot}=5000) and the corresponding cycle number N_{cyc} are marked at the top-left corner, respectively. It can be seen that (i) all loops are well enclosed; (ii) for N_{cyc} altering from 1 to 9, the damping ratios are very low since their surfaces are extremely minor (pseudo-elastic); (iii) from N_{cyc} equalling to 13, the hysteresis loops become more obvious with the significant increase in D; (iv) in the stage of large deformation (e.g., N_{cyc}=24, 27), the two corresponding damping ratios attain the high level of 55.79% and 59.66%, reconfirming the occurrence of liquefaction softening.

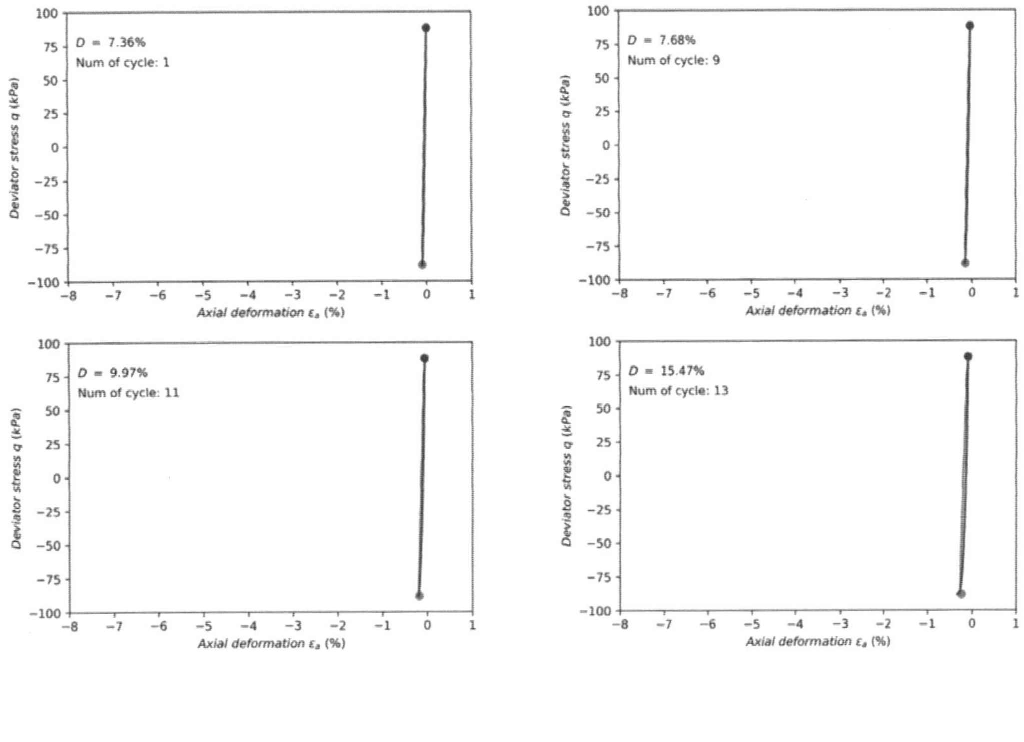

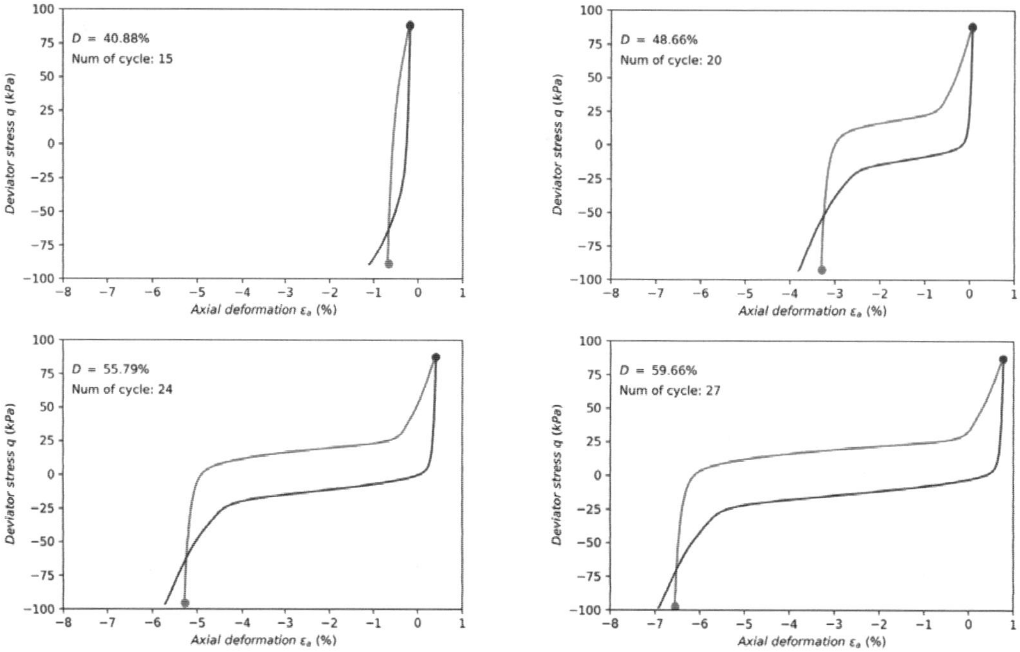

Figure. 4 Treated hysteresis loops and corresponding damping ratios of HN31 sand using Monte-Carlo method (T_{ot}=5000)

4. Reliability

Figure. **5** presents damping ratio D against deviator strain ε_d curve. Note that in undrained triaxial condition, ε_d is equal to ε_a by combining the following two factors: (i) $\varepsilon_d=2(\varepsilon_a-\varepsilon_r)/3$ and (ii) $\varepsilon_v=(\varepsilon_a+2\varepsilon_r)/3=0$, where $\varepsilon_a/\varepsilon_r$ stand for axial/radial principal stresses and ε_v represents volumetric strain. In the case of T_{ot}=25, the damping curve is characterized by a fake peak value of about 82%, which is contradictory to the sand behavior identified in Figure. **3**. This misleading could be attributed to the fact that the random points are too few (only 25) to gain an accurate probability related to the damping ratio. In the cases of T_{ot}=500, 5000, 10000 and 50000, it is found that (i) the fake peak damping ratio disappears; (ii) with the increase in T_{ot}, the curves gradually converge to a single one, especially for T_{ot}=5000, 10000 and 50000; (iii) three distinct responses of sand specimens could be clearly identified, which is consistent with experimental results in Figure. **3**. While assessing damping ratio using Monte-Carlo method, the above phenomena further suggest that (i) T_{ot} is the most prominent factor and (ii) a value higher than 5000 times could, in general, ensure a satisfactory accuracy.

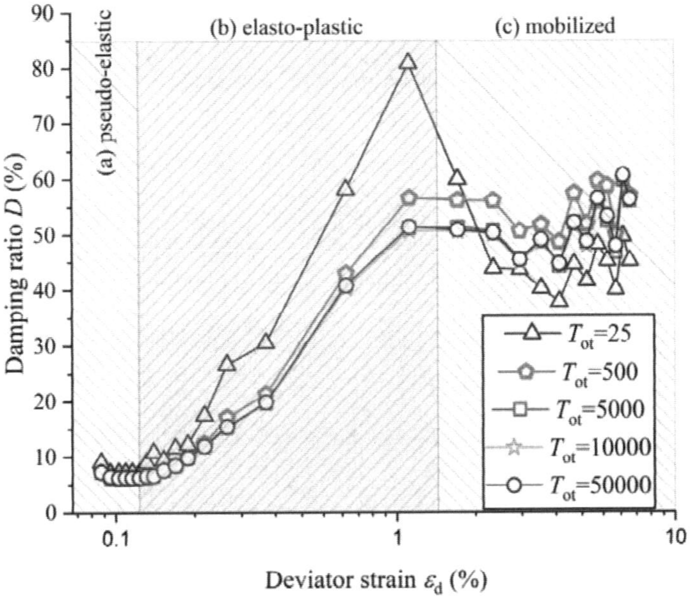

Figure. 5 Evolution of damping ratio D in terms of total random samplings number T_{ot} for each hysteresis loop

5. Conclusion
This paper investigates an innovant method for determining the damping ratio with Monte-Carlo method. Compared with other approaches presented in the literature, the great advantage of the method is directly treating irregular hysteresis loops without any geometrical assumption. In addition, it is found that the computed result is highly sensitive to T_{ot} and a value of T_{ot} higher than 5000 times could, in general, provide an accurate result.

Acknowledgement
The financial support provided by the research sponsors through ANR ISOLATE and the priceless helps provided by Miss Li are gratefully acknowledged. The Python program (version: 3.7.7) employed in this study is available upon requesting.

Reference
[1] Sun J, Wu G, Zhang J, Zeng Y and Xiao W 2017 Experimental study on concrete columns reinforced by hybrid steel-fiber reinforced polymer (FRP) bars under horizontal cyclic loading *Constr. Build. Mater.* **130** 202–211
[2] Li W, Lang L, Wang D, Wu Y and Li F 2018 Investigation on the dynamic shear modulus and damping ratio of steel slag sand mixtures *Constr. Build. Mater.* **162** 170–180
[3] Zhu Z, Kham M, Fernandes V and Lopez-Caballero F 2020 Dynamic Response of a Central Clay Core Dam Under Two-Component Seismic Loading *Springer Ser. Geomech. Geoengin.* **1** 231–236
[4] Kramer S 1996 *Geotechnical Earthquake Engineering* (Person Education India)
[5] Wei M and Lin M 1993 A new method for determining the equivalent damping ratio of soil *Water Resour. Hydropower. Eng.* 50–53
[6] Chen W, Kong L and Zhu J 2007 Simple method to approximately determine the damping ratio of soils Rock Soil Mech 28 789–791
[7] Benahmed N 2001 Mechanical behaviour of a sand under monotonic and cyclic loading: application

to liquefaction and cyclic mobility phenomena Ph.D Thesis Ecole Nationale Des Ponts et Chaussées
[8] Zhu Z, Dupla J, Canou J and Foerster E 2020 Experimental study of liquefaction resistance : effect of non-plastic silt content on sand matrix Eur. J. Environ. Civ. Eng. 0 1–19
[9] Jradi L 2018 Study of the influence of fine particles on the properties of liquefaction of sands Ph.D Thesis University Paris-Est

IOP Conf. Series: Materials Science and Engineering 1048 (2021) 012004 doi:10.1088/1757-899X/1048/1/012004

Photocatalytic treatment of water contaminated by organic dye with ETS-10 titanium silicate

A Cantiello[1], S Candamano[2], P De Luca[2]*

[1]Dipartimento di Ingegneria per l'Ambiente ed il Territorio ed Ingegneria Chimica. Università della Calabria, Arcavacata di Rende (Cs). I-87036 –Italy.
[2]Dipartimento di Ingegneria Meccanica, Energetica e Gestionale. Università della Calabria, Arcavacata di Rende (Cs). I-87036 Rende- Italy.

*E-mail: pierantonio.deluca@unical.it

Abstract. In this research, ETS-10 titanium silicate (Engelhard titanium silicate) was studied for the treatment of water contaminated by the organic dye called Blu Patent V, which is commonly used in the textile industries. In addition to being a microporous material, the ETS-10 phase also has photocatalytic capacities. In particular, predetermined quantities of ETS-10 have been added to the Blu Patent V solutions and the entire system has been exposed, for programmed times, to sunlight. Systems also in the presence of hydrogen peroxide were used. The resulting solutions were analyzed with UV spectrophotometry to evaluate the reduction of the dye. ETS-10, after treatment, was analyzed by thermogravimetric analysis (TG).

1. Introduction

Water pollution [1] is a deeply felt problem at the environmental level as well as air pollution [2] and soil [3].The quality and healthiness of the water is put at risk by a multitude of pollutants. These can be inorganic elements such as heavy metals [4-13]. The organic substances that can be found in the waters are many and varied and come from anthropic and industrial activities [14-19]. All these substances can be particularly dangerous for human health and can also be found within the tissues and organs [20-25]. The water analysis and the recognition of polluting agents are carried out through various chemical-physical techniques such as nuclear magnetic resonance, spectroscopic techniques, thermal analysis on residues etc. [26-36]. The problem of water pollution is also addressed through the improvement of the sectors that provide for the use of environmentally sustainable substances and with a low emission of polluting substances in the environment [37-42]. There are many researches that aim to enhance and seek innovative treatments for water treatment. The use of microporous materials such as zeolites [43-46] or carbon nanotubes [14, 47-49] has proven to be particularly effective. The latter are very versatile materials and can be used as adsorbent materials and as fiber reinforcing [50, 51] Photodegradation techniques, which use photocatalytic agents and light are inexpensive and very effective [52]. The purpose of this work is to use a particular material which in addition to being microporous material, and which therefore can perform an adsorbent action, also has a proven photocatalytic activity. The material used in this research was the ETS-10 (Engelhard titanium silicate) molecular sieves. The used ETS-10 phase is in sodium-potassium form and can give adsorption or ion-exchange reactions, as shown in the following simplified expression: (Na, K) ETS-10 + $M^{+2n} \rightleftarrows$ (M)ETS-10 + Na^{+} + K^{+} [53-58]. The latter has been used for the treatment of water contaminated by the dye, Blue Patent V, generally used in the industrial sector and is often present in waste water.

2. Experimental

The ETS-10 phase used in this research has been previously synthesized in accordance with the procedures reported in reference [46]. The organic dye used was Blu Patent V. It is a dark blue synthetic dye. The appearance is of a very water soluble violet powder [52]. The concentration of the

Blu Pantent V solution used was equal to 0.1g / L. The systems used for experimentation are shown below:

System 1: 150 ml Blu Patent V solution + 1 g of ETS-10

System 2: 150 ml Blu Patent V + 1 g of ETS-10 + 10 ml H_2O_2 (30%VV)

The systems thus prepared were exposed for different times to direct sunlight. At time intervals of 4, 8, 12, 16, 20 days, the systems were filtered by a vacuum pump and the solutions obtained were analyzed by a UV spectrophotometer. The solid residues were analyzed by thermogravimetric analysis (TG).

3. Results

The following figure 1 shows the concentrations of the Blue Patent V solutions with ETS-10 (system 1) after programmed times of exposure to sunlight. It is possible to note that only after 8 days of exposure to sunlight there is a reduction of over 50%. After 20 days it reaches close to 90% reduction.

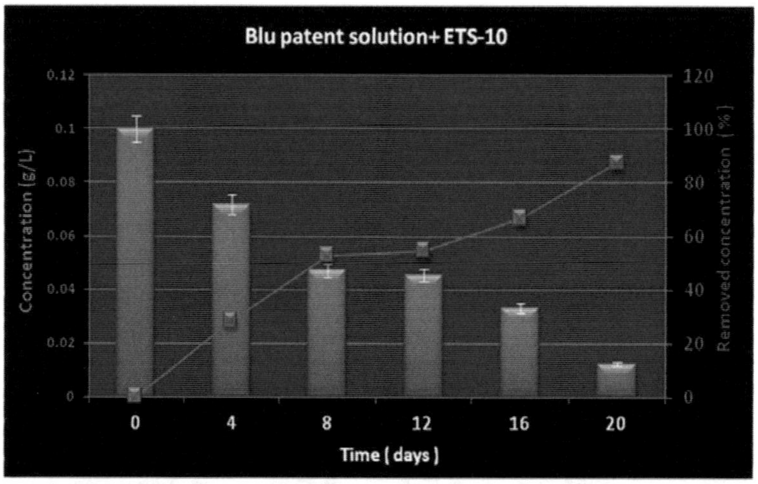

Days	0	4	8	12	16	20
Concentration (g/l)	0.10	0.071	0.047	0.045	0.033	0.012
Removed concentration (%)	0	28.15	53.00	54.67	66.80	87.50

Figure. 1 Variation of the concentration of the Blu Patent V dye in system 1 as a function of the time of exposure to sunlight

Figure 2 shows the concentrations of the Blu Patent V solution in the presence of ETS-10 and hydrogen peroxide (system 2) as a function of the time of exposure to sunlight. It is possible to notice how with the passing of the days there is a decrease of the dye due to the contemporary presence of ETS-10 and H2O2. An important reduction of over 90% respect to initial concentration is reached after four days of exposure to sunlight. The following Figure 3a shows the TG analysis of the ETS-10 phase before and after the treatments. The results obtained show that no weight variation of the ETS-10 before and after the treatments is evident. This shows that the ETS-10 phase did not act as an adsorbent but as a photocatalytic material. Figures 3(b) e (c) show the degradation of the solution color, respectively for 1 and 2 systems, at different exposure times.

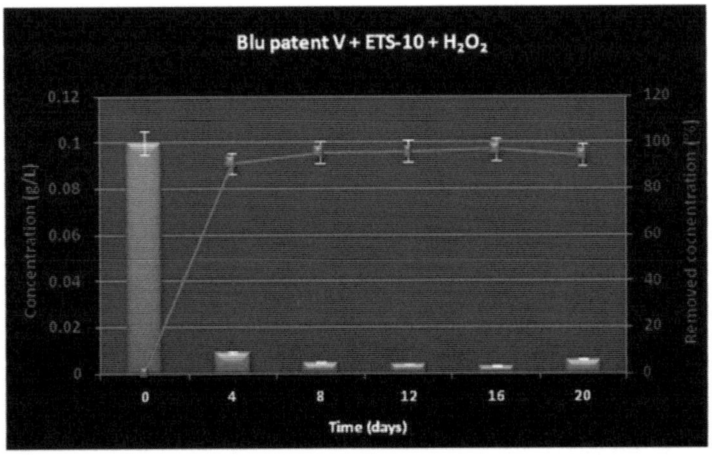

Days	0	4	8	12	16	20
Concentration (g/L)	0.10	0.009	0.0045	0.004	0.0031	0.0056
Removed concentration (%)	0	90.96	95.43	96.04	96.90	94.36

Figure. 2 Variation in the concentration of the patent V blue dye in system 1 as a function of the time of exposure to sunlight

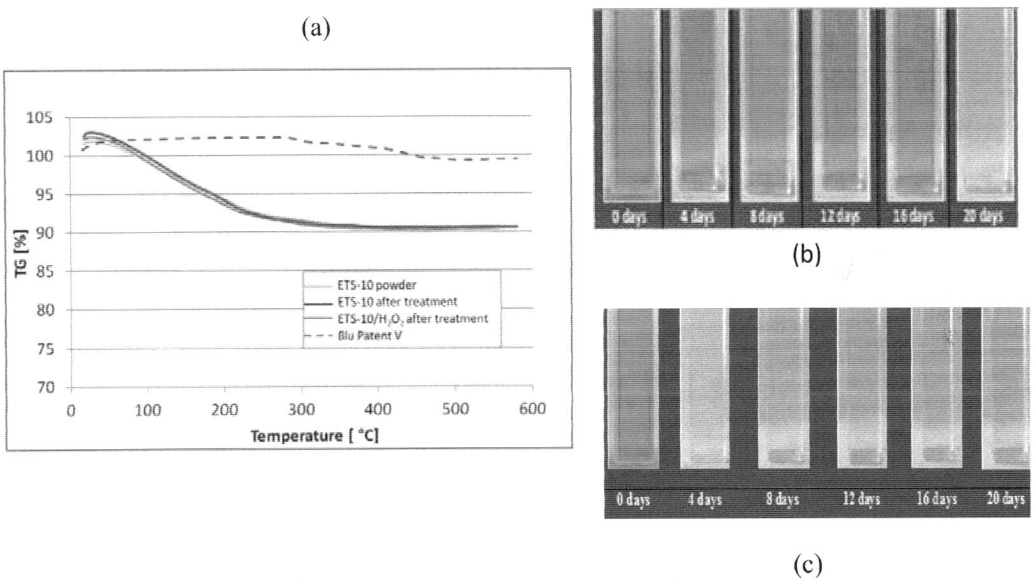

Figure. 3 (a)Thermogravimetric analysis (TG) on the ETS-10 phase in powder before of the treatment, and after the treatments without and with hydrogen peroxide (system 1 and 2 respectively) and Blu Patent V; Coloring of solutions as a function of exposure time to sunlight for system 1 (b) and system 2 (c)

Conclusion

The preliminary results obtained showed that the ETS-10 phase allows the reduction of the Blu Patent V dye by exposure to sunlight, reaching a reduction of approximately 50% after eight days. The addition of hydrogen peroxide accelerates the dye abatement process in fact after only 4 days, about 90% abatement is achieved. In the systems studied, the ETS-10 phase acted as a photocatalitic material and not as an adsorbent agent, probably due to the large size of the dye molecule. The results are particularly encouraging for the simplicity and cost-effectiveness of the treatment..

References

[1] Ranieri E, Bombardelli F, Gikas P, Chiaia B. *Applied and Environmental Soil Sciencen* 2016, Article ID 9415175, http://dx.doi.org/10.1155/2016/9415175
[2] Filice M, De Luca P, Guido G P 2009. *Environ Eng Manag J* 8 (**6**) 1407-1412
[3] Malandrino M, Abollino O, Buoso S, Giacomino A, La Gioia C, Mentasti E 2011. *Chemosphere* **82**:169–178
[4] De Raffele G, Aloise A, De Luca P, Vuono D, Tagarelli A, B.Nagy J 2016. *J. Porous Mater.* 23 (**2**) 389-400.
[5] Materazzi S, Risoluti R, Napoli A 2015 *Thermochim. Acta* **606** 90-94
[6] Aiello D, Furia E, Siciliano C, Bongiorno D, Napoli A, 2018 *J Mol Liq* **269** 387-397
[7] Risoluti R, Piazzese D, Napoli A, Materazzi S, 2016 *J. Anal. Appl. Pyrolysis* **117** 82-87
[8] Materazzi S, Foti C, Crea F, Risoluti R, Finamore J. 2014 *Thermochim. Acta* **580** 7–12
[9] Vecchio S, Materazzi S, Wo LW, De Angelis Curtis S 2013 *Thermochim. Acta* **568** 31-37
[10] Materazzi S, Napoli A, Risoluti R, Finamore J, D'Arienzo S 2014 *Int. J. Mass Spectrom.* **365-366** 372-376.
[11] Materazzi S, Finamore J, Risoluti R, Napoli A 2014 *Microchemical Journal* **115** 27-31.
[12] Kurdziel K, Glowiak T, Materazzi S, Jezierska J, 2003 *Polyhedron* **22** 3123-3128
[13] Catauro M, Naviglio D, Risoluti R, Vecchio Ciprioti S 2018 *J. Therm. Anal. Calorimetry* **133** 1085–1092.
[14] Lico D, Vuono D, Siciliano C, B.Nagy J, De Luca P 2019. *J. Environ. Manag.* **237** 636–643.
[15] Cardiano P, Falcone G, Foti C, Giuffrè O, Napoli A, 2013 *Journal of Inorganic Biochemistry* **129** 84-93.
[16] De Luca P, Foglia P, Siciliano C, B.Nagy J, Macario, A. 2019 *Environments* – 6 (**9**) Article number 101
[17] Temperini A, Aiello D, Mazzotti F, Athanassopoulos C M, De Luca P, Siciliano C 2020 *Molecules* **25 (6)** Article number 25061313
[18] Reddy CE, Albanito L, De Marco P, Aiello D, Maggiolini M, Napoli A, Musti AM 2013 *Cell Death and Disease*, **4 (10)** e852
[19] Aiello D, Cardiano P, Cigala RM, Gans P, Giacobello F, Giuffrè O, Napoli A, Sammartano S 2017 *J Chem Eng Data* **62 (11)** 3981-3990
[20] Di Donna L, Benabdelkamel H, Taverna D, Indelicato S, Aiello D, Napoli A, Sindona G, Mazzotti F 2015 *Anal. Bioanal. Chem*, **407 (19)**,5835-5842.
[21] Sergi M, Gentili A, Perret D, Marchese S, Materazzi S, Curini R. *Chromatographia* **65** 757-761
[22] Aiello D, Siciliano C, Mazzotti F, Di Donna L, Athanassopoulos CM, Napoli A, 2020 *Food Chem* **307** 125527
[23] Aiello D, Casadonte F, Terracciano R, Damiano R, Savino R, Sindona G, Napoli A, 2016 *Oncoscience* **3 (7-8)** 220-241
[24] Marchetti L, Sabbieti MG, Menghi M, Materazzi S, Hurley MM, Menghi G 2002 *Histology and Histopathology* **17** 1061-1066
[25] Aiello D, Materazzi S, Risoluti R, Thangavel H, Di Donna L, Mazzotti F, Casadonte F, Siciliano C, Sindona G, Napoli A 2015 *Mol. BioSyst.* **11** 2373-2382
[26] Materazzi S, Risoluti R. 2014 *Appl. Spectrosc. Rev.* **49** 635-665
[27] Cardiano P, Giuffrè O, Napoli A, Sammartano S, 2009 *New J Chem* **33 (11)** 2286-2295
[28] Napoli A, Aiello D, Aiello G, Cappello MS, Di Donna L, Mazzotti F, Materazzi S, Fiorillo M, Sindona G 2014 *J. Prot. Res.* **13 (6)** 2856-2866
[29] Jahouh F, Saksena R, Aiello D, Napoli A, Sindona G, Kováč P, Banoub JH, 2010 *J Mass Spectrom* **45 (10)** 1148-1159
[30] Aiello D, Siciliano C, Mazzotti F, Di Donna L, Athanassopoulos CM, Napoli A, 2018 *RSC Advances* **8 (63)** 36104-36113

[31] Materazzi S, De Angelis Curtis S, Vecchio Ciprioti S, Risoluti R, Finamore J, 2014 *J. Therm. Anal. Calorim.* 116 93–98
[32] Napoli A, Athanassopoulos CM, Moschidis P, Aiello D, Di Donna L, Mazzotti F, Sindona G, 2010 *Anal Chem* **82** **(13)** 5552-5560
[33] Romolo FS, Ferri E, Mirasoli M, D'Elia M, Ripani L, Peluso G, Risoluti R, Maiolini E, Girotti, S.2015 *Forensic Science Internaltional* **246** 25-30.
[34] Risoluti R, Gregori A, Schiavone S, Materazzi S 2018 *Anal. Chem.* 90 4288-4292
[35] Materazzi S, Peluso G, Ripani L, Risoluti R 2017 *Microchem. J.* 134 277-283
[36] Mazzotti F, Benabdelkamel H, Donna LD, Athanassopoulos CM, Napoli A, Sindona G 2012 *J. Mass Spectrom* **47** **(7)** 932-939
[37] De Luca P, Carbone I, B.Nagy J 2017. *J. Green Build.* 12 **(4)** 141–161.
[38] Nastro V, Vuono D, Guzzo M, Niceforo G, Bruno I, De Luca P 2006 *J Therm Anal Calorim* 84 **(1)** 181-184
[39] De Luca P, De Luca P, Candamano S, Macario A, Crea F, B.Nagy J. 2018. *Buildings*, 8**(9)** 122
[40] Navarra MA, Materazzi S, Panero S, Scrosati B 2003 *J Electrochem Soc* **150** A1528-A1532
[41] De Luca P, Chiodo A, B.Nagy J 2011 *Sustain. Chem.* **154** 155–165.
[42] De Luca P, Roberto B, Vuono D, Siciliano C, B.Nagy J. 2018. *IOP Conf. Ser. Mater. Sci. Eng.* 374 **(1)** 012071. doi:10.1088/1757-899X/374/1/012071
[43] Veltri M, Vuono D, De Luca P, B.Nagy J, Nastro A 2006. *J Therm Anal Calorim* 84 **(1)** 247-252
[44] De Luca P, Crea F, Fonseca A, B.Nagy, J 2001 *Microporous and Mesoporous Materials* 42 **(1)** 37-48
[45] De Luca P, Bernaudo I, Elliani R, Tagarelli A, B.Nagy J, Macario A 2018 *Materials* 11 **(11)** 2316.
[46] De Luca P, Poulsen T G, Salituro A, Tedeschi A, Vuono D, Kònya Z, Madaràsz D, B.Nagy J 2015. *J. Therm. Anal. Calorim.* 122 **(3)** 1257–1267.
[47] Foglia P, Vuono D, Siciliano C, Napoli A, B.Nagy J, De Luca P. 2019. *IOP Conf. Ser. Mater. Sci. Eng.* 572 **(1)** 1,Article number 012047
[48] Policicchio A; Vuono D, Rugiero T, De Luca P, B.Nagy, J 2015. J. CO2 Util **10** 30–39.
[49] De Benedetto C, Macario A, Siciliano C, B.Nagy J. 2020 Materials 13 **(12)** 2757; doi:10.3390/ma13122757
[50] De Luca P, Pane L, Vuono D, Siciliano C, Candamano S, B.Nagy J 2017. *Environ. Eng. Manag. J.* 16 **(8)** 1659–1672.
[51] De Luca P, Nappo G, Siciliano C, B.Nagy J 2018. *J. Porous Mater.* 25**(1)** 283–296
[52] Cantiello A, Candamano S, De Luca P. 2020. *IOP Conf. Ser. Mater. Sci. Eng.* 739 **(1)** Article number 012054
[53] Vuono D, Guzzo M, De Luca P, B.Nagy J 2014. *J. Therm. Anal. Calorim.* 116 **(1)** 169–182
[54] De Luca P, Vuono D, Filice M 2009. *Environ. Eng. Manag. J* 8 **(5)** 1009-1015
[55] De Luca P, Mastroianni C, B.Nagy J. 2018 *IOP Conf. Ser. Mater. Sci. Eng* 374 **(1)** Article number 012003
[56] Turta N A, De Luca P, Bilba N, B.Nagy J, Nastro A 2008 *Microp Mesopor Mat* 112 **(1-3)** 425-431
[57] Vuono D, De Luca P, B.Nagy J, Nastro A 2008. *Microp Mesopor Mat* 109 **(1-3)** 118-137
[58] Pavel C C, Vuono D, De Luca P, Bilba N, B.Nagy J, Nastro A 2005 *Microp Mesopor Mat* 80 **(1-3)** 263-268

IOP Conf. Series: Materials Science and Engineering 1048 (2021) 012005 doi:10.1088/1757-899X/1048/1/012005

Experimental measurements through ultrasounds for viscoelasticity analysis

A Maiorano[1], G Napolitano Dell'Annunziata[1] and E Rocca[1]

[1]Department of Industrial Engineering, University Federico II, Naples, Italy,

E-mail: antonio.maiorano@unina.it;

Abstract. Among the different methodologies useful to determinate polymers viscoelastic properties in literature, it is possible to find some different techniques based on the ultrasounds pulse-echo methods which show several limits in the properties estimation of the tested materials. In this scenario, the proposed manuscript focuses on an innovative non-destructive technique of viscoelastic materials by means of ultrasounds, which could be suitably employed on polymers. The developed procedure is aimed to evaluate the material response in terms of loss factor within different temperatures and induced excitation frequencies. To carry out the viscoelasticity analysis through ultrasounds, a bench has been set up and the data acquisition has been performed by means of the pulse-echo method on three different tires tread specimens and, then, a post-process algorithm has been implemented in order to determine the viscoelastic behavior in terms of loss factor. The comparison of the achieved results with the temperature sweep master curves, which were determined by means of standard Dynamic Mechanical Analysis (D.M.A.) testing on the same polymers, has provided an indication of the goodness of the proposed procedure.

1. Introduction

The evaluation of viscoelasticity is a fundamental topic in a wide range of activities concerning the development of polymers for innovative compounds, the parametrization of physical contact models [1,2] and the evaluation of the grip coefficient [3,4] for the optimization of vehicle performance and road safety. For these applications, the estimation of experimental data concerning the viscoelasticity of the compound can be further useful for the definition of mathematical models [5,6] describing properly the dynamic behavior in every conditions.

The Dynamic Mechanical Analysis (D.M.A.) is widely employed into the characterization of viscoelasticity in order define the hysteretic behavior of the specimen following the Time-Temperature Superposition principle. This technique requires testing a specimen of suitable dimensions within a frequency range from 0.1 Hz to 100 Hz by means of different rheometers [7,8]. However, in peculiar cases and circumstances, in which a specific geometry should be extracted to realize a particular specimen compound, this methodology cannot be employed and it is necessary to develop a novel innovative and non-destructive methodology to overcome these limits and to characterize the material viscoelastic properties.

In this scenario, the analysis of the ultrasonic wave propagation and attenuation in the medium is one of the most efficient and simplest technique for the viscoelastic characterization in a wide frequency and temperature domain [9,10]. Particularly, the transmitter mode is widely used for the evaluation of the viscoelastic properties of the specimen of interest by means of the analysis of the F.F.T. performed on both the transmitter and received signals [11–13]. However, this approach is not suitable in many mechanical systems consisting of a mixing of polymer with other type of materials, such a passenger or a motorsport tire. On the other hand, the pulse-echo method, which requires the usage of a single probe as transmitter and receiver, can be simply employed into the analysis of these complex systems analyzing only the reflected waves of interest at each interface [14,15].

Therefore, this manuscript proposes a methodology for the evaluation of the viscoelastic properties through the ultrasounds as function of temperature and frequency. For this reason, an experimental session on three tire tread compounds has been carried out at the ultrasound bench of the Tire Lab of the University of Naples Federico II in order to measure the magnitudes to be processed for the viscoelasticity characterization, using the pulse-echo method.

2. Viscoelasticity theory

A generic viscoelastic material exhibits mechanical properties that vary with time and temperature. Particularly, it halfway behaves between a purely elastic and a purely viscous one. Thus, during a standard D.M.A., it is possible to observe a phase delay between stress and the corresponding strain compared towards time, as shown in Figure 1. This means that the stress-strain relationship is defined by a complex dynamic modulus as amount of the overall resistance to deformation of the compound [16,17]:

$$\frac{\sigma(\omega)}{\varepsilon(\omega)} = E^* = E' + E'' \tag{1}$$

where the real part, E', is the Storage Modulus and the imaginary part, E'', is the Loss Modulus. These magnitudes are deeply related with the way the material dissipates a part of energy provided by means of a load/stress time function. The phase angle δ depends on the Storage and Loss Modulus as follows:

$$\tan(\delta) = \frac{E'}{E'} \tag{2}$$

The magnitudes in eq. 2 are function of the solicitation frequency and the material temperature. Taking into account the solicitation as the wave pressure at a certain frequency, the loss factor can be expressed as:

$$\tan(\delta) = \frac{E'}{E'} = \frac{2(a \cdot c/\omega)}{1-(a \cdot c/\omega)^2} \tag{3}$$

where α is the attenuation coefficient, expressed in dB/mm, and it is a measure of the amplitude damping phenomenon occurring during the wave propagation in a dissipative medium, c is the sound velocity in the medium and ω is the angular pulsation of the propagating wave [11,13]. For this reason, to determine the loss factor, the authors have carried out experimental measurements through ultrasounds useful to evaluate the viscoelastic properties of specimen of vulcanized compound employing the pulse-echo method.

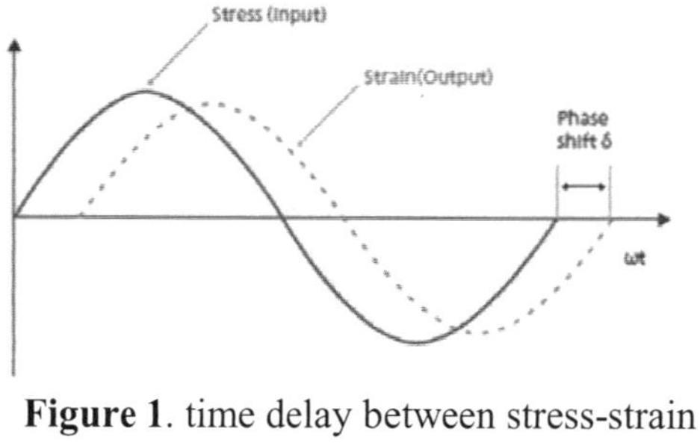

Figure 1. time delay between stress-strain

3. Experimental measurements

The ultrasound experiments have been carried out on tire compounds in the Tire Lab of Department of Industrial Engineering at the University of Naples Federico II. To perform the underwater ultrasonic tests, an ultrasound bench was settled, as shown in Figure. 2. This bench involves a water tank, in which the compound samples are immersed. The water has an appropriate acoustic impedance and a suitable sound attenuation coefficient at different temperature, for these reasons has been selected as coupling fluid for the presented test activity. A pair of electrical heating resistances have been set in order to warm up tire compounds and the water contained in the tank. In addition, a pump is adopted for water circulation in order to maintain a homogeneous temperature in each point of the tank, while a thermostat checks the tire temperature by means of a thermocouple, setting the water and thus the compound temperature in the tank.

To generate ultrasonic waves, an echograph and a pair of immersion probes have been employed. The first instrument is used to generate electrical signal in the range of 60 - 200 V that, thanks to piezoelectricity phenomenon, is converted into pressure wave. Particularly, as the transmitter voltage value rises, the acoustic power associated with the incident wave and its maximum amplitude will be higher.

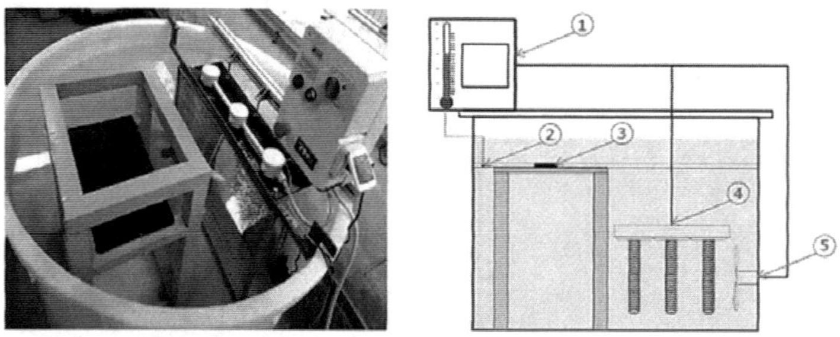

Figure 2. Ultrasounds test rig (on the left); Rig conceptual scheme (on the right)

Each probe generates longitudinal ultrasounds at a specific frequency, 0.4 MHz and 1.2 MHz, respectively. In the experimental activity proposed, three specimens of vulcanized polymers, denoted as compound C1, C2 and C3, have been tested. The employment of slabs is advantageous for their flat surface. For each specimen, the test temperature range is approximately 15-60 degrees due to immersive probe working limit conditions. As mentioned before, the pulse-echo measurement method has been performed. On the echograph display, the user can analyze the amplitude values of the backward reflected waves in a decibel scale according to the A-Scan representation. As shown in Figure 3, the maximum amplitude values of the front echo, A_1, and its multiple reflection, A_1', can be measured by means of the A-scan available on the echograph display.

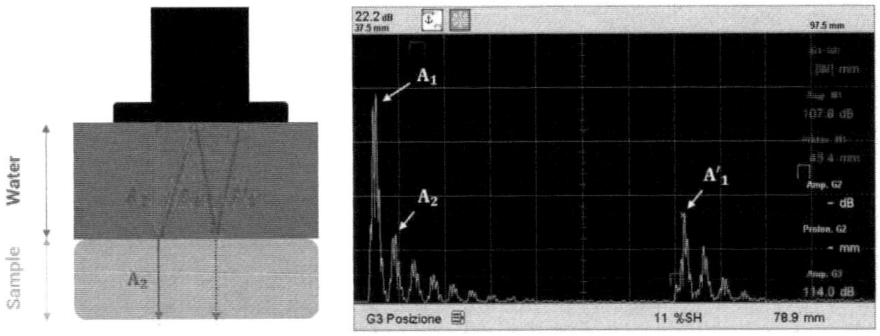

Figure 3. Pulse-echo mode outline (on the left); A-Scan representation (on the right)

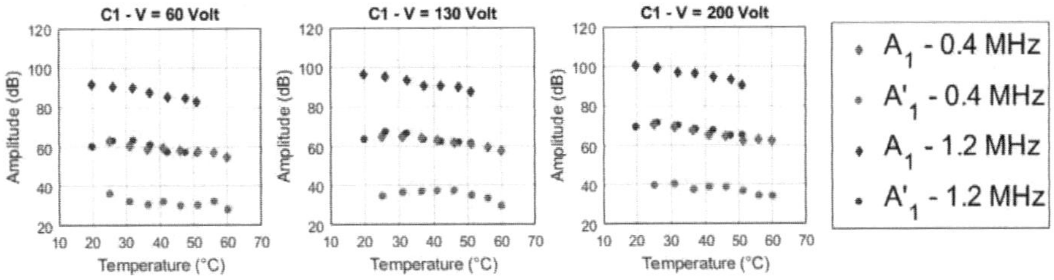

Figure 4. Measured amplitudes as function of temperature and frequency for compound C1

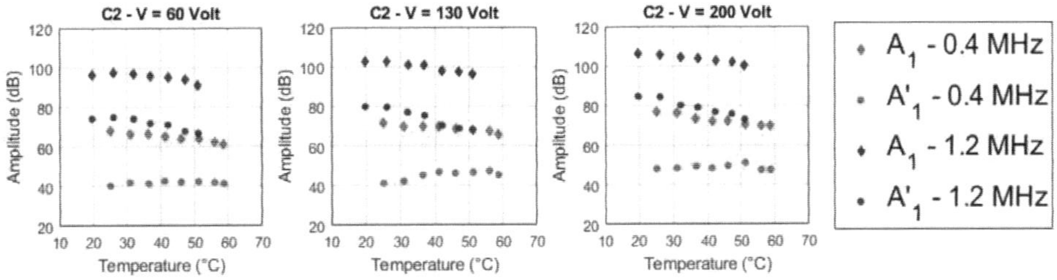

Figure 5. Measured amplitudes as function of temperature and frequency for compound C2

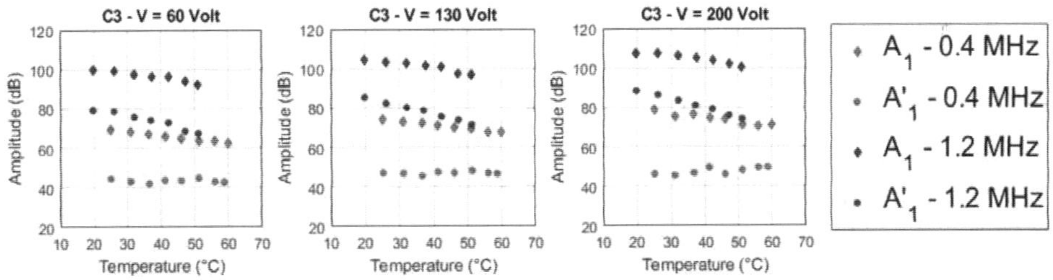

Figure 6. Measured amplitudes as function of temperature and frequency for compound C3

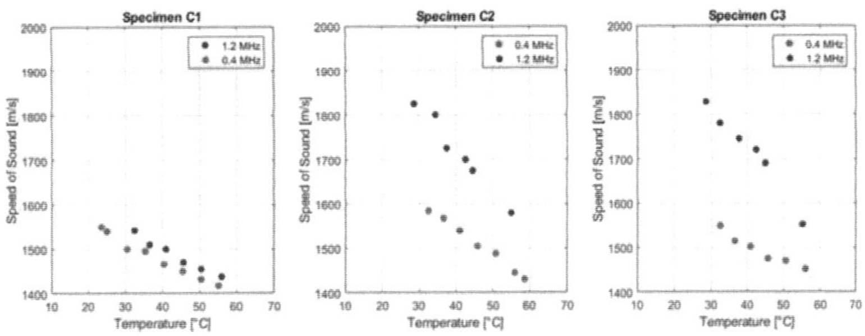

Figure 7. Measured sound velocity values for each specimen

The acquired values for each tested compound are diagrammed in Figures 4, 5 and 6. As noticeable, the measures have been carried out taking into account 5 transmitter voltages for each specimen in order to store a proper amount of data for the viscoelasticity analysis in every condition. The sound velocity towards the temperature have been measured for each specimen employing both probes: the front echo amplitudes and the sound velocity values exhibit a downward trend as the temperature increases (see Figure 7). This means that the material properties, as well as the acoustic ones, change over the temperature due to the weakening of the polymer chains.

4. Ultrasounds data processing

The ultrasound experimental data acquired on viscoelastic samples have been used to characterize the compound acoustic properties and, then, the loss factor. Particularly, taking into account the measured amplitudes and sound velocity values, the wave damping coefficient within the medium has been determined optimizing the following cost function [18]:

$$\varepsilon = \left[\left(A_{1_{DATA}} - A_{1_{Eval}}\right);\left(A'_{1_{DATA}} - A'_{1_{EVAL}}\right)\right] \tag{4}$$

where $A_{1_{DATA}}$ and $A'_{1_{DATA}}$ are the data acquired during the experimental test session; $A_{1_{Eval}}$ and $A'_{1_{EVAL}}$ are evaluated in the proposed algorithm as follows:

$$A_{1_{Eval}} = A_0 \cdot D_w \cdot |R^*_{w\,c}| \cdot |T_{w\,p}| \tag{5.a}$$

$$A'_{1_{EVAL}} = A_0 \cdot D_w^2 \cdot |R^*_{w\,c}|^2 \cdot |T_{w\,p}| \cdot |R_{w\,p}| \tag{5.b}$$

where A_0 is the transmitted pulse, $T_{w\,p}$ is the transmission coefficient at the water-probe interface, and $R_{w\,p}$ is the reflection coefficient at the water-probe interface. These magnitudes depend on the transmitter voltage and temperature and they are provided by the probe manufacturer. $|R^*_{w\,c}|$ is the modulus of the complex reflection coefficient at water-compound interface:

$$R^*_{w\,c} = \frac{Z_c^* - Z_w}{Z_c^* + Z_w} \tag{6}$$

where Z_w is the water acoustic impedance and Z_c^* is the complex acoustic impedance of the viscoelastic specimen, which depends on the unknown attenuation coefficient, compound density, sound velocity and frequency [13].

The results of the identified attenuation coefficient for each compound by means of the optimization of the cost function in eq. 4 are shown in Figure 8.

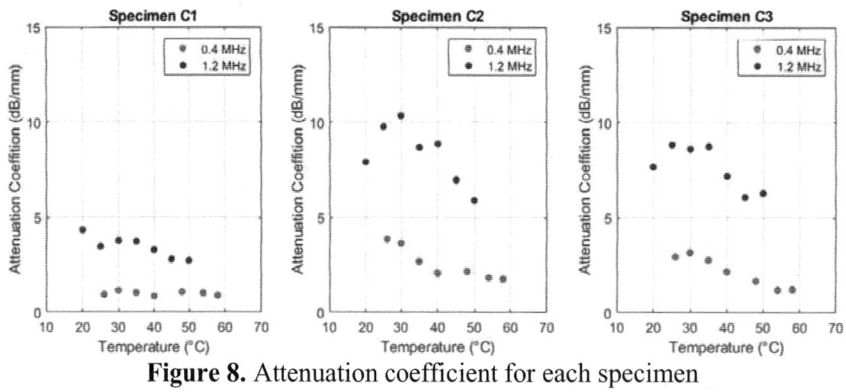

Figure 8. Attenuation coefficient for each specimen

5. Loss Factor evaluation

In order to evaluate the loss factor values at 0.4 and 1.2 MHz as function of temperature, as shown in eq. 3, for each different compound specimen it is necessary to consider the density values, the sound velocity and the estimated attenuation coefficient. The viscoelastic index values thus obtained refer to the temperature corresponding to the test experimental sessions and are in according to the working limit conditions of the specific immersive probe used. To verify the trustworthiness of the identified viscoelastic loss factor, the comparison of these values with the master curve performed in temperature sweep at 1 Hz in a cantilever clamp is necessary. Particularly, the William-Landel-Ferry relationship is applied in order to evaluate the temperature shifting in according to the following formula [19,20]:

$$\log_{10} a_T = \frac{-C_1(T - T_0)}{C_2 + (T - T_0)} \tag{7}$$

where the fitting constant C_1 and C_2 have been identified by means of D.M.A. frequency sweep test at different temperatures on the same compound specimen. The temperature shifting from the starting value corresponding to 0.4 MHz and 1.2 MHz to the equivalent ones at the frequency of interest (1 Hz) has led to the comparison of the ultrasound tests acquired data with the D.M.A. master curve, as shown in Figure 9.

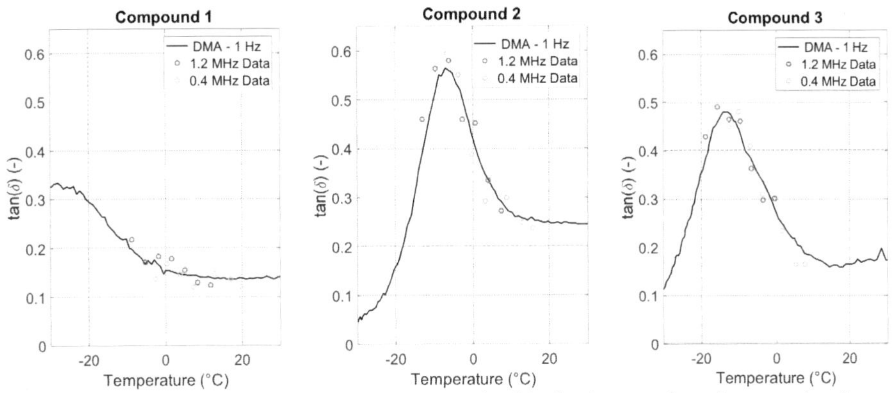

Figure 9. D.M.A. 1 Hz master curves compared with the innovative characterization results

It is noticeable how the acquired data with the innovative ultrasound methodology are in good agreement with the D.M.A. temperature sweep master curves in the shifted temperature range. Particularly, higher the frequency of the ultrasounds, lower the minimum shifted temperature: thus, the analysis can provide an indication of the glass transition temperature of the compound, especially for the compound 2 and compound 3.

6. Conclusions

In this manuscript an innovative and non-destructive methodology involving ultrasound technology, developed with the aim to evaluate the viscoelastic properties of compound polymers has been presented. The main novelty of this work is the develop of an identification algorithm able to evaluate the attenuation coefficient depending on temperature and frequency variations considering the experimental data relative to the front echo and its multiple reflection at water-specimen interface acquired during a test session through the pulse-echo transmitter mode for ultrasound analysis.

In order to carry out the experimental test, a dedicated ultrasound bench test has been designed and set up to test different compound specimens. These slabs have been tested using two different ultrasonic probes varying the temperature. The results obtained with the described technique have been compared with the standard master curve obtained by means of the D.M.A. tests with the aim to validate the procedure and evaluate the goodness of the loss factor identification.

The trustworthiness of the achieved results can be considered as baseline for the viscoelastic materials properties evaluation through non-destructive technique. Further analysis will be focused on testing employing ultrasonic waves in a wide frequency spectrum in order to characterize the compounds viscoelastic behavior within the entire temperature range of interest.

References

[1] Genovese A, Farroni F, Papangelo A and Ciavarella M 2019 A Discussion on Present Theories of Rubber Friction, with Particular Reference to Different Possible Choices of Arbitrary Roughness Cutoff Parameters *Lubricants* **7** 85

[2] Persson B N J 2001 Theory of rubber friction and contact mechanics *J. Chem. Phys.* **115** 3840–61

[3] Ciaravola V, Farroni F, … G F-E and 2017 U *engineeringletters.com*

[4] Genovese A, Ciavarella M, Di Bari P, Farroni F, Papangelo A, Genovese A, Carputo F, Ciavarella M, Farroni F, Papangelo A, Sakhnevych A, Carputo F and Sakhnevych A 2019 Analysis of Multiscale Theories for Viscoelastic Rubber Friction MegaRide-applied vehicle research-an UniNa spin-off company View project Local friction of rubber with rough surfaces View project ANALYSIS OF MULTISCALE THEORIES FOR VISCOELASTIC RUB-BER FRICTION *Springer* 1125–35

[5] Genovese A, Carputo F, Maiorano A, Timpone F, Farroni F and Sakhnevych A 2020 Study on the Generalized Formulations with the Aim to Reproduce the Viscoelastic Dynamic Behavior of Polymers *Appl. Sci.* **10** 2321

[6] Sasso M, Palmieri G and Amodio D 2011 Application of fractional derivative models in linear viscoelastic problems *Mech. Time-Dependent Mater.* **15** 367–87

[7] Lakes R S 2009 *Viscoelastic materials* (Cambridge University Press)

[8] Lakes R S 2004 Viscoelastic measurement techniques *Rev. Sci. Instrum.* **75** 797–810

[9] Krautkrämer J and Krautkrämer H 1990 *Ultrasonic Testing of Materials* (Springer Berlin Heidelberg)

[10] Oprlsonl C A, Alshuth T and Schuster R H 2008 Polymer dynamics at high frequencies in dependence of the polymer structure *KGK Kautschuk Gummi Kunststoffe* **61** 301–5

[11] Lionetto F and Maffezzoli A 2009 Polymer characterization by ultrasonic wave propagation *Adv. Polym. Technol.* **27** 63–73

[12] Lionetto F, Montagna F and Maffezzoli A 2005 Ultrasonic dynamic mechanical analysis of polymers *Appl. Rheol.* **15** 326–35

[13] Mc Hugh J 2008 Ultrasound Technique for the Dynamic Mechanical Analysis (DMA) of Polymers. 63–4

[14] Carlson J E, Van Deventer J, Scolan A and Carlander C 2003 Frequency and temperature dependence of acoustic properties of polymers used in pulse-echo systems *Proc. IEEE Ultrason. Symp.* **1** 885–8

[15] Gerspacher M, O'Farrell C P, Nikiel L, Yang H H and Le Méhauté F 1996 *High frequency viscoelasticity of carbon black filled compounds* vol 69 (Rubber Division of the American Chemical Society)

[16] Ferry J 1980 Viscoelastic properties of polymers

[17] Mark J E, Erman B and Roland M 2013 *The science and technology of rubber*

[18] Waltz R A, Morales J L, Nocedal J and Orban D 2006 An interior algorithm for nonlinear optimization that combines line search and trust region steps *Math. Program.*

[19] Williams M L, Landel R F and Ferry J D 1955 The Temperature Dependence of Relaxation Mechanisms in Amorphous Polymers and Other Glass-forming Liquids *J. Am. Chem. Soc.* **77** 3701–7

[20] Landel R F 2006 A Two-Part Tale: The WLF Equation and Beyond Linear Viscoelasticity *Rubber Chem. Technol.* **79** 381–401

IOP Conf. Series: Materials Science and Engineering 1048 (2021) 012006 doi:10.1088/1757-899X/1048/1/012006

Complex formation of copper(II), nickel(II) and γ-L-glutamyl-L-cysteinyl-glycine (GSH).

Napoli Anna

Dipartimento di Chimica e Tecnologie Chimiche. Università della Calabria, Arcavacata di Rende (Cs). I-87036 –Italy.

E-mail: amc.napoli@unical.it

Abstract. The formation of Cu(II) and Ni(II) complexes of GSH in conditions of 2-fold GSH excess over Ni(II) was studied by MALDI MS and MS/MS. The instrumental data demonstrated that *S* and *N* of GSH are involved in the coordination metal spheres, while COOH did not participate to complexation. NiL_2 is a square-planar Ni(II) specie able to form one electron redox couple with Ni- (III) complex. It was found to dimerize rapidly via disulphide bond formation in the presence of air.

1. Introduction

Anthropogenic activities [1-5] have led to high emissions of toxic metals into the environment [6.7]. Glutathione (γ-L-glutamyl-L-cysteinyl-glycine, GSH) is a ubiquitous tripeptide found in the cells of plants, animals and microorganisms. It plays a crucial role in protecting intracellular components against oxidative damage and detoxifying heavy metal ions in an organism [8]. This tripeptide bearing two free -COOH groups [9] and a NH_2 group, provides a hydrophilic interface and a handle for further reactivity with other functional molecules as well as metal ions. The purpose of our study was not only evaluating metal speciation by mass spectrometry [10] but also to obtain, isolate and characterize GSH-M(II) complexes. Interest towards Cu and Ni-GSH systems arises from an autoxidative processes that has been observed with nickel and copper complexes of XXH-CO_2H peptide. Over the years, spectroscopic techniques [11-15] have become a powerful tool to identify [16-21] and quantitate [22,23] metabolites [24-27], amino acids, proteins [28-31] and their post-translational modifications [32,33]. Electrospray mass spectrometry (ESI-MS) has been used for the detection and analysis of metal ion complexes [34-44]. Matrix assisted laser desorption mass spectrometry (MALDI-TOF/TOF-MS) is more generally used for investigations of the most types of large biomolecules [45,46]. It is characterized by short analysis time, high sensitivity, tolerance to contaminants and the ability to detect different components in a mixture [47,48]. The analysis of low molecular weight analytes, such as drugs and metabolites, can be performed by tandem mass spectrometry (MS/MS) that provides the required sensitivity and molecular specificity In this work, MALDI-TOF/TOF-MS was used to investigate the composition and elucidate the structures of complexes of GSH with Cu(II) and Ni(II) cations.

2. Materials and Methods

Materials. GSH, $NiCl_2$, Cu (SO_4) and α-cyano-4-hydroxy-trans-cynnamic acid (α-CHCA, 99.0%) are purchased from Sigma Aldrich- Merck (Milano, Italy).

Synthesis of M^{2+}(Cu, Ni) –glutathione complexes.

Complexes were synthesized according to previously published procedures [49-52]. Briefly, all [M(GSH)$_2$] complexes were prepared by adding the GSH (2 mmol) dissolved in water to a magnetically stirred aqueous solution containing the appropriate metal (1 mmol). The resulting mixtures were kept at room temperature for 3h. The coordination compounds were isolate by treatment with ethanol. The resulting precipitate was dissolved in water. A 1 µL portion of a premixed solution of each sample and α-CHCA (0.3% in TFA) was spotted on the matrix target, dried at room temperature and directly analyzed by MALDI mass spectrometry.

Mass spectrometry

MS and MS/MS analyses were performed using a 5800 MALDI TOF-TOF Analyzer (AB-SCIEX) equipped with a neodymium-yttrium-aluminum-garnet laser (laser wavelength 349 nm). At least 4000 laser shots were typically accumulated with a laser pulse rate of 400 Hz in the in reflection positive-ion mode with a mass accuracy of 5 ppm, whereas in the MS/MS mode spectra up to 5000 laser shots were acquired and averaged with a pulse rate of 1000 Hz. MS/MS experiments were performed with a mass accuracy of 50 ppm and at a collision energy of 1 kV, and ambient air was used as the collision gas with a medium pressure of 10−6 Torr. After acquisition, spectra were handled using Data Explorer version 4.0.

3. Results and discussion

The ion-providing information on ML_n (n: 1, 2) species was observed in the full-scan positive ion mass spectra of Cu:GSH (**1**) and Ni:GSH(**2**). MS information combined with data collected by MS/MS at 1 KeV of collision energy were used to assign the most probable structures of the complexes and the coordination sites.

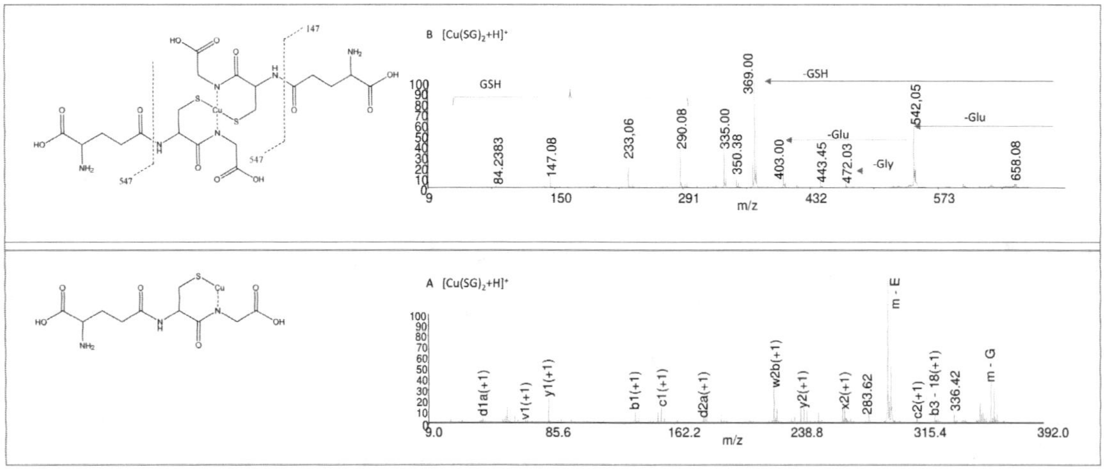

Figure. 1 MALDI MS/MS spectrum of (A) [Cu(SG)2-H]+ (**1**, m/z 676.0915), (B) [Cu(SG)-H]$^+$ (**2**, m/z 369.0065)

MALDI MS spectra obtained for **1** reveals the species [Cu(SG)-H]$^+$ ([$C_{10}H_{16}CuN_3O_6S$]$^+$, m/z 369.0065, Δppm =4), arising from the decomposition laser induced of [Cu(SG)$_2$-H]$^+$, and the [Cu(SG)$_2$-H]$^+$ of m/z 676.0915 ([$C_{20}H_{33}CuN_6O_{12}S_2$]$^+$, Δppm =4). The [Cu(SG)$_2$+H]$^+$ ion was decomposed to various fragment ions by loss of small molecules, such as water ([Cu(SG)$_2$+H–H_2O]$^+$ of m/z 658.08 ([$C_{20}H_{31}CuN_6O_{11}S_2$]$^+$) and by cleavage of peptide bonds leading to loss of glutamic acid moiety [Cu(SG)$_2$+H–Glu]$^+$ (m/z

547.06, $[C_{15}H_{26}CuN_5O_9S_2]^+$), $[Cu(SG)_2+H\text{-}2Glu]^+$ (m/z 403.00, $[C_{10}H_{18}CuN_3O_6S_2]^+$) and $[Cu(SG)_2+H\text{-}GluGly]^+$ (m/z 472.03, $[C_{13}H_{21}CuN_4O_7S_2]^+$), similar to the MS/MS spectrum of $[Cu(SG)+H]^+$ and noncomplexed glutathione (Figure 1A-B). These results confirm that CuL_2 contains two bidentate GSH molecules, with the Cu(II) environment of S2N2 (Figure1). Reaction of $NiCl_2$ with GSH (**2**) was of particular interest because thiol group undergoes spontaneous reaction after complexation of nickel(II). Direct MS analysis of solutions containing both Ni(II) species and ligand with a final M:L ratio of 1 :2 showed the formation of the ions of m/z 613.1617 ($[C_{20}H_{33}N_6O_{12}S_2]^+$), m/z 635.1437 ($[C_{20}H_{32}N_6NaO_{12}S_2]^+$), m/z 650.0631 ($[C_{20}H_{28}N_6NiO_{11}S_2]^+$), m/z 668.0738 ($[C_{20}H_{30}N_6NiO_{12}S_2]^+$) and m/z 671,0973 ($[C_{20}H_{33}N_6NiO_{12}S_2]^+$). The ion of m/z 613 was assigned to the oxidized glutathione arising from a Ni(II)-based redox process. The formation of the ion species $[GSSGNi]^+$ and $[GSSGNi+H\text{-}H_2O]^+$ suggested that the oxidation product (GSSG) is a good ligand for Ni(II) binding itself. The complexes of m/z 668 and m/z 650 arise from aerobic oxidation of Ni(II)-GSH complex. MS/MS experiments of the m/z 668 ([GSSG]) show ion products arising by the cleavage of the peptide bond via loss of Gly (m/z 593) and Glu (m/z 539) residues, respectively (Figure 2). The detection of the ions of 308 ($[GSH+H]^+$ and its counterpart (m/z 363), due to the cleavage of disulfide bond, indicates that the ion specie of 668 contains two bidentate GSH molecules and the coordination occurs *via* N and COOH. MS/MS experiments clearly indicate that $[Ni(GS)_2+H]^+$ contains two bidentate GSH molecules, with the Ni(II) environment of S2N2. According literature data, NiL_2 (**2**) is a square-planar Ni(II) specie able to form one electron redox couple with Ni- (III) complex[53].

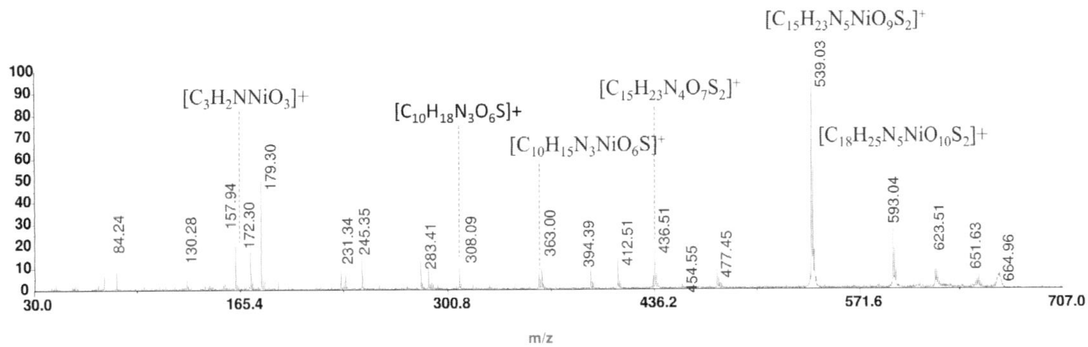

Figure. 2 MALDI MS/MS Spectrum of m/z 668 [Ni(GSSG)].

Conclusion

We described for the complexation of GSH with the environmental relevant cations Cu(II) and Ni(II). Under the adopted conditions, ML and ML_2 mononuclear complexes were prevalent, and the formation of polymetallated species was not observed along all experiments. The instrumental data demonstrated that *S* and *N* of ligand are involved in the coordination metal spheres, while COOH did not participate to complexation. The square-planar Ni(II) specie $[Ni(GS)_2]$ is susceptible to air oxidation.

Reference

[1] De Luca P, Carbone I, Nagy J B *Journal of Green Building*141–161, 12, 2017.
[2] Filice M, De Luca P, Guido G P *Environmental Engineering and Management Journal*, 1407-1412, 8, 2009.
[3]. Vuono D, De Luca P, Nagy J B, Nastro A *Microporous and Mesoporous Materials*, 118-137, 109, 2008.
[4] Nastro V, Vuono D, Guzzo M, Niceforo G, Bruno I, De Luca P *Journal of Thermal Analysis and Calorimetry*, 247-252, 84, 2006.
[5] Pavel C C, Vuono D, De Luca P, Bilba N, Nagy J B, Nastro A *Microporous and Mesoporous Materials*, 263-268, 80,2005.

[6] De Luca P, Nappo G, Siciliano C, Nagy J B *Journal of Porous Materials* 283-296, 25, 2018.
[7] De Luca P, Pane L, Vuono D, Siciliano C, Candamano S, Nagy J B *Environmental Engineering and Management Journal*, 1659-1672, 16, 2017.
[8] Marchetti L, Sabbieti M G, Menghi M, Materazzi S, Hurley M M, Menghi G *Histology and Histopathology*, 1061-1066, 17, 2002.
[9] Materazzi S, Foti C, Crea F, Risoluti R, Finamore J *Thermochimica Acta*, 7-12, 580, 2014.
[10] Cardiano P, Falcone G, Foti C, Giuffrè O, Napoli A *Journal of Inorganic Biochemistry*, 84-93, 129, 2013.
[11] Risoluti R, Fabiano M A, Gullifa G, Vecchio Ciprioti S, Materazzi S *Applied Spectroscopy Reviews*, 39-72, 52, 2017.
[12] Navarra M A, Materazzi S, Panero S, Scrosati B *Journal of the Electrochemical Society*, A1528-A1532, 150, 2003.
[13] Materazzi S, Peluso G, Ripani L, Risoluti R *Microchemical Journal*, 277-283, 134, 2017.
[14] Materazzi S, De Angelis Curtis S, Vecchio Ciprioti S, Risoluti R, Finamore J *Journal of Thermal Analysis and Calorimetry*, 93-98, 116, 2014.
[15] Risoluti R, Gullifa G, Battistini A, Materazzi S *Analytical Chemistry*, 6435-6439, 91(10), 2019.
[16] Risoluti R, Gregori A, Schiavone S, Materazzi S *Analytical Chemistry*, 4288-4292, 90, 2018.
[17] Risoluti R, Caprari P, Gullifa G, Massimi S, Sorrentino F, Buiarelli F, Materazzi S *Microchemical Journal*, 374-380, 146, 2019.
[18] Materazzi S, Risoluti R, Pinci S, Romolo F S *Talanta*, 673-678, 174, 2017.
[19] Risoluti R, Materazzi S, Sorrentino F, Bozzi C, Caprari P *Talanta*, 216-222, 183, 2018.
[20] Risoluti R, Gullifa G, Fabiano M A, Sorrentino F, Caprari P, Materazzi S *Journal of Thermal Analysis and Calorimetry*,1299-1306,134, 2018.
[21] De Luca P, Pane L, Vuono D, Siciliano C, Candamano S, Nagy J B *Environmental Engineering and Management Journal*, 1659–1671, 16, 2017.
[22] Mazzotti F, Benabdelkamel H, Di Donna L, Athanassopoulos C M, Napoli A, Sindona G *Journal of Mass Spectrometry* 932-939, 47, 2012.
[23] Napoli A, Athanassopoulos C M, Moschidis P, Aiello D, Di Donna L, Mazzotti F, Sindona G *Analytical Chemistry* 5552-5560 82 (13), 2010.
[24] Romolo F S, Ferri E, Mirasoli M, D' Elia M, Ripani L, Peluso G, Risoluti R, Maiolini E, Girotti S *Forensic Science International*, 25-30, 246, 2015.
[25] Sergi M, Gentili A, Perret D, Marchese S, Materazzi S, Curini R *Chromatographia* 757-76, 65, 2007.
[26] Gentili A, Caretti F, D'Ascenzo G, Mainero Rocca L, Marchese S, Materazzi S, Perret D *Chromatographia* 669-676, 66, 2007.
[27] Risoluti R, Gullifa G, Battistini A, Materazzi S *Drug and Alcohol Dependence*,107578, 205, 2019.
[28] Aiello D, Casadonte F, Terracciano R, Damiano R, Savino R, Sindona G, Napoli A *Oncoscience* 220-224, 3 (7-8), 2016.
[29] Napoli A, Aiello D, Aiello G, Cappello M S, Di Donna L, Mazzotti F, Materazzi S, Fiorillo M, Sindona G *Journal of Proteome Research*, 2856-2866, 13(6), 2014.
[30] Imbrogno S, Aiello D, Filice M, Leo S, Mazza R, Cerra M C, Napoli A *Scientific Reports*, 9(1),18953, 2019.
[31] Agas D, Marchetti L, Menghi G, Materazzi S, Materazzi G, Capacchietti M, Hurley M M, Sabbieti M G *Journal of Cellular Physiology*, 145-152, 214, 2008.
[32] Jahouh F, Saksena R, Aiello D, Napoli A, Sindona G, Kováč P, Banoub J H *Journal of Mass Spectrometry*, 1148-1159, 45(10), 2010.
[33] Reddy C E, Albanito L, De Marco P, Aiello D, Maggiolini M, Napoli A, Musti A M *Cell Death and Disease*, 4 (10), art. no. e852, 2013DOI: 10.1038/cddis.2013.381.
[34] Aiello D, Cardiano P, Cigala R M, Gans P, Giacobello F, Giuffrè O, Napoli A, Sammartano S *Journal of Chemical and Engineering Data*, 3981-3990, 62(11), 2017.
[35] Materazzi S, Risoluti R, Napoli A *Thermochimica Acta*, 90-94, 606, 2015.

[36] Cardiano P, Giuffre O, Napoli A, Sammartano S *New Journal of Chemistry*, 2286-2295, 33(11), 2009.
[37] Aiello D, Furia E, Siciliano C, Bongiorno D, Napoli A *Journal of Molecular Liquids*, 387-397, 269, 2018.
[38] Furia E, Aiello D, Di Donna L, Mazzotti F, Tagarelli A, Thangavel H, Napoli A, Sindona G *Dalton Transactions,* 1055-1062, 43 (3), 2014.
[39] Materazzi S, Vecchio S, Wo L W, De Angelis Curtis S *Thermochimica Acta*, 183-187, 543, 2012.
[40] Beneduci A, Corrente G A, Marino T, Aiello D, Bartella L, Di Donna L, Napoli A, Russo N, Romeo I, Furia E Journal *of Molecular Liquids* 111805, 296, 2019.
[41] Vecchio S, Materazzi S, Wo L W, De Angelis Curtis S *Thermochimica Acta*, 31-37, 568, 2013.
[42] Papadopoulos C, Cristvo B, Ferenc W, Hatzidimitriou A, Vecchio Ciprioti S, Risoluti R, Lalia-Kantouri M *Journal of Thermal Analysis and Calorimetry* 717–729, 123, 2016.
[43] Materazzi S, Finamore J, Risoluti R, Napoli A *Microchemical Journal*, 27-31, 115, 2014.
[44] Kurdziel K, Glowiak T, Materazzi S, Jezierska J *Polyhedron* 3123-3128, 22, 2003.
[45] Materazzi S, Napoli A, Risoluti R, Finamore J, D'Arienzo S *International Journal of Mass Spectrometry*, 372-376, 365-366, 2014.
[46] Salvatore L, Gallo N, Aiello D, Lunetti P, Barca A, Blasi L, Madaghiele M, Bettini S, Giancane G, Hasan M, Borovkov V, Natali M L, Campa L, Valli L, Capobianco L, Napoli A, Sannino A *International Journal of Biological Macromolecules* 291-306, 154, 2020
[47] Aiello D, Siciliano C, Mazzotti F, Di Donna L, Athanassopoulos C M, Napoli A RSC *Advances* 36104-3611, 38 (63), 2018
[48] Aiello D, Siciliano C, Mazzotti F, Di Donna L, Athanassopoulos C M, Napoli A *Food Chemistry* 125527, 307, 2020.
[49] Catauro M, Naviglio D, Risoluti R, Vecchio Ciprioti S Journal *of Thermal Analysis and Calorimetry,* 1085–1092, 133, 2018.
[50] Migliorati V, Ballirano P, Gontrani L, Materazzi S, Ceccacci F, Caminiti R *Journal of Physical Chemistry B*, 7806-7818, 117, 2013.
[51] Giuffrè O, Aiello D, Chillè D, Napoli A, Foti C *Environmental Science: Processes and Impacts*, 1731-1742, 22, 2020.
[52] Chillè D, Aiello D, Grasso GI, Giuffrè O, Napoli A. Sgarlata C, Foti C, *Journal of Environmental Sciences (China),* 100-110, 94, 2020.
[53] Ross S A, Burrows C J *Inorganic Chemistry,* 5358–5363, 37, 1998.

The use of Hydrazine-based Derivatization technique to characterize glycosyl esters of crocetin from Saffron.

Donatella Aiello

Dipartimento di Chimica e Tecnologie Chimiche. Università della Calabria, Arcavacata di Rende (Cs). I-87036 –Italy.

E-mail: donatella.aiello@unical.it

Abstract. Saffron is one of the oldest natural dyestuffs and is obtained from the dried stigmata of *Crocus sativus* L. The yellowness of saffron results from the presence of crocins (glycosyl esters of crocetin). In this report we present the identification and characterization of multiple forms of crocins from whole saline saffron extracts using direct MALDI MS and MS/MS of glycosyl esters of crocetin and of their corresponding Hydrazine tagged products.

1. Introduction

Saffron, the red dried stigmas of Crocus sativus L., belongs to Iridaceae family and is the world's most expensive spice. The major components of saffron are the apocarotenoids cis- and trans-crocins, responsible for the typical yellow-gold colour [1-3]. The saffron colorants were identified in various sample matrices and determined by many spectroscopic methods (table 1). Chromatography is recommended as the most efficient technique for saffron analysis, in particular high performance liquid chromatography with UV–Vis/ ESI-MS detection [4-10]. Over the years, spectroscopic methods [11-20] have become powerful analytical platforms [21-33] to identify and quantitate metabolites [34-39], amino acids [40-41], proteins [42-46] and their post-translational modifications [47,48]. In particular, MALDI mass spectrometry, for its high specificity and better resolution with respect to a conventional chromatographic approach, represents the methodology of choice to get reliable results in the profiling of natural matrices [49].

Table1. Glycosyl esters of crocetin identified in saffron extract by LC-UV-Vis spectra and LC/MS.

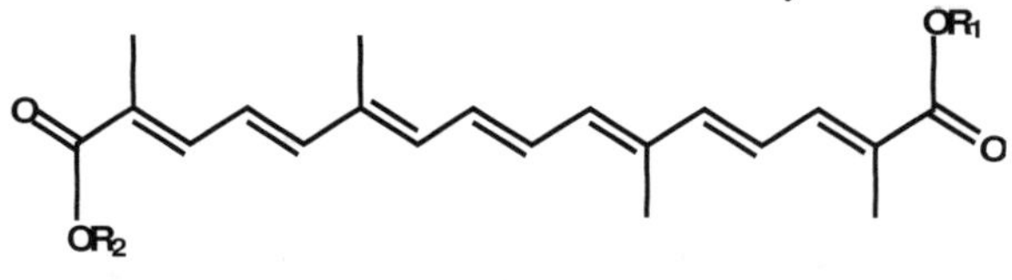

Name	R_1	R_2	Formula	MW
Crocin-5 (C5)	Three β-D-glucosyl	β-D-gentobiosyl	$C_{50}H_{74}O_{29}$	1138.4
Crocin-4 (C4)	β-D-gentobiosyl	β-D-gentobiosyl	$C_{44}H_{64}O_{24}$	976.4
Crocin-3 (C3)	β-D-gentobiosyl	β-D-glucosyl	$C_{38}H_{54}O_{19}$	814.3
Crocin-2 (C2)	β-D-gentobiosyl	H	$C_{32}H_{44}O_{14}$	652.3
Crocin-2' (C2)	β-D-glucosyl	β-D-glucosyl	$C_{32}H_{44}O_{14}$	652.3
Crocin-1 (C1)	β-D-glucosyl	H	$C_{26}H_{34}O_9$	490.2
Crocetin (C)	H	H	$C_{20}H_{24}O_4$	328.2

Moreover, there is a growing tendency to find rapid, simple and high-throughput strategies to characterize complex samples Chemical derivatization, combined with mass spectrometry, offers several opportunities [50]. In particular hydrazine reagents are a well-known group of derivatizing agents for the determination of aldehydes. The hydrazine tagged products are stable, compatible with mass spectrometry analysis and the reaction is fast and quantitative. Therefore, we propose a simple method to obtain hydrazine tagged products useful to study MS/MS fragmentations pathways suitable for a simple characterization of natural glycated species in saffron, focusing our attention towards glycosyl esters of crocetin.

2. Materials and Methods

Saffron sample preparation. The dried saffron stigmas were purchased from Cooperative of Saffron, Krokos Kozanis, Greece. Crocin solutions were obtained from 5 mg portions of dried saffron stigmas, and extracted and treated as reported [3]. The crude extract was directly analysed by MALDI MS and MS/MS. To obtain sugar oxidation, crocins desalting was performed using solid phase extraction (C18 cartridges) and the eluted fraction was collected and dried in vacuum using Centrivap Concentrator. Dry sample was dissolved in 200μL of oxidation buffer (0.1M sodium acetate buffer, pH 5.5), added with 1mg of sodium-meta-periodate and gently stirred for 2 hours at room temperature. 1μL of crocin mixture containing reactive aldehydes was added with 1μL of a hydrazine solution (14mM, in sodium acetate buffer, pH 5) and 4 μL of α-CHCA matrix (5mg/mL). The sample solution (0.5μL) was loaded on to MALDI plate and allowed to air-dry.

Mass spectrometry analysis. MALDI-TOF analyses were performed using a 5800 MALDI TOF/TOF analyzer (AB SCIEX), in reflectron positive mode, 4000 laser shots were typically accumulated, with a laser pulse rate of 400 Hz. For MALDI-TOF/TOF analysis, collision induced dissociation (CID) was performed at collision energy of 1 kV, and ambient air was used as the collision gas with a medium pressure of 10^{-6} Torr, accumulating 5000 laser shots.

3. Results and discussions

The commercially available stigmas of Greek Crocus Sativus cultivars were selected as case studies to determine the crocins profile of the whole extract and to identify and characterize specific crocins without any previous chromatographic separation. The high solubility in aqueous medium is an important prerequisite for organoleptic, colouring and taste saffron properties. The partition coefficient of crocins and the profile of saffron are strongly related to the solvents used for the extraction. An ammonium bicarbonate solution was chosen to prepare the whole crocin extract from saffron of Greek cultivars.

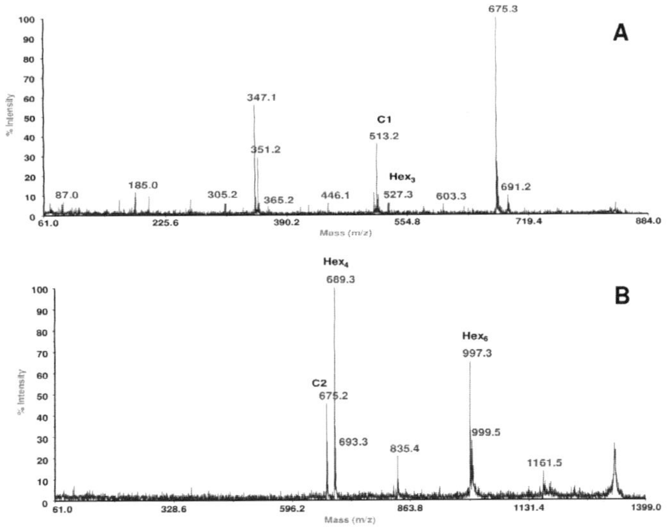

Figure 1. MALDI MS/MS spectra of [C3+Na]+, m/z 837.3 (A) and [C6+Na]+, m/z 1323.4 (B).

Ammonium bicarbonate should help the formation of glycosyl/ammonium ion pairs, thus favouring the solubility of the crocins to be extracted in moderately polar solvents. Moreover, ammonium counter-ions have been often used to improve the desorption of high molecular weight proteins and do not interfere with the mass spectrometric analysis. The crude extract was directly analysed by MALDI TOF in the reflectron mode. MALDI MS profiling of the whole extract allowed the identification of the crocetin derivatives bearing from 6 to 9 glycosyl units [1,3]. MALDI MS/MS experiments were performed, aiming to elucidate the glycosylation composition of crocins (figure 1).

CROCINS	Sodiated MS/MS fragments						Hypothetical structures
Crocin 8	C6; m/z 1323.5	C5; m/z 1161.4	Hex6; m/z 997.3	Hex4; m/z 689.3	C2; m/z 675.3		Hex6-Crocetin-Hex2
Crocin 7	C6; m/z 1323.5	C5; m/z 1161.4	Hex6; m/z 997.3	Hex5; m/z 835.3	Hex4; m/z 689.3		Hex6-Crocetin-Hex1
Crocin 6	C5; m/z 1161.4	C4; m/z 999.4	Hex6; m/z 997.3	Hex4; m/z 689.3	C2; m/z 675.3		Hex6-Crocetin Hex4-Crocetin-Hex2
Crocin 5	C4; m/z 999.4	C3; m/z 837.3	Hex4; m/z 689.3	C2; m/z 675.3			Hex3-Crocetin-Hex2 Hex4-Crocetin-Hex1
Crocin 4	C3; m/z 837.3	C2; m/z 675.3	Hex2; m/z 365.3	Crocetin; m/z 351.3	Crocetin –CO2; m/z 305.1		Hex2-Crocetin-Hex2
Crocin 3	C2; m/z 675.3	Hex3; m/z 527.3	C1; m/z 513.2	Hex2; m/z 365.3	Crocetin; m/z 351.3	Crocetin –CO2; m/z 305.1	Hex3-Crocetin Hex2-Crocetin-Hex1
Crocin 2	C1; m/z 513.2	Hex2; m/z 365.3	Crocetin; m/z 351.3	Crocetin –CO2; m/z 305.1			Hex2-Crocetin

Table 2. Sodiated MS/MS fragments and hypothetical structures of crocins.

It has been established clearly that esters undergo three main MS/MS fragmentation reactions involving the formation of (a) the acylium ion, (b) the alkyl carbonium ion and (c) the alcohol. Crocetin glycosyl esters, under CID conditions, show only two different fragmentation pathways: (1) acylcarbon-oxygen bond cleavage yielding the intact glycan of the highest MW followed by the competitive formation of the corresponding crocin counterpart; (2) cross ring fragmentation of the glycan moiety, characterized by cleavage of glycosidic bonds as well as a 0,2X-ring fragmentation of the outermost unit. Based on the observed fragmentation pattern can be presumed that, **C3** (m/z 837.3, figure 1A) represents a mixture of Hex3-crocetin and Hex2-crocetin-Hex1, while **C6** (m/z 1323.4, figure 1B) could be a mixture of Hex6-crocetin and Hex4-crocetin-Hex2. All MS/MS data and crocins hypothetical structures are reported in Table 2. Since carbonyl function has higher ionization efficiency than carbohydrates, glycans are not cleaved specifically, so the MS/MS data allow the determination of unclear crocins structures.

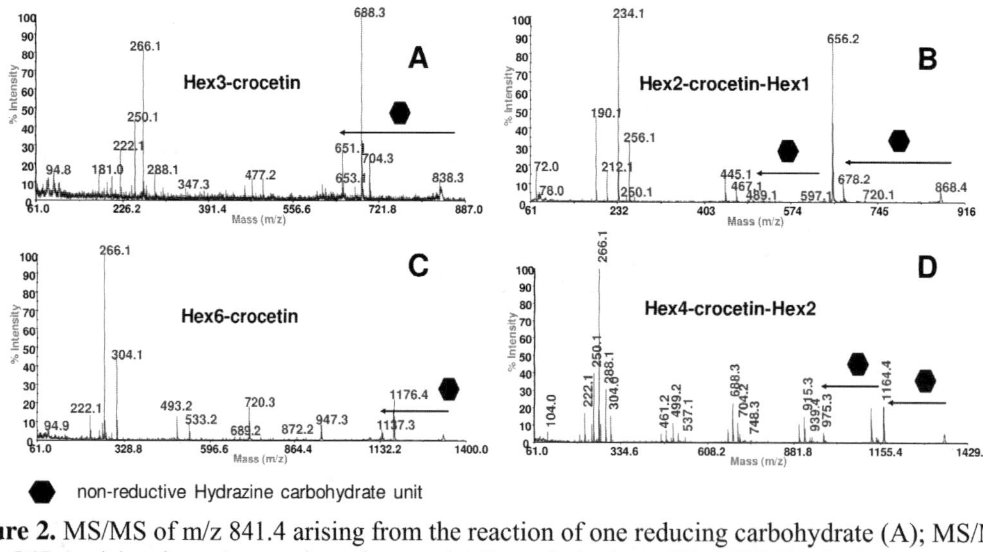

Figure 2. MS/MS of m/z 841.4 arising from the reaction of one reducing carbohydrate (A); MS/MS of m/z 867.4 arising from the reaction of two reducing carbohydrate (B); MS/MS of m/z 1327.5 arising from the reaction of two reducing carbohydrate (C); MS/MS of m/z 1353.6 arising from the reaction of two reducing carbohydrate (D).

The derivatization of the reducing end of saccharides can influence sensitivity, ionization efficiency and fragmentation behaviour. Therefore, the number of glycan fragments can be enhanced on going from non-reductive Hydrazine tagged to glycosyl esters of crocins. Crocetin glycosyl esters were derivatized with NH_2NH_2 after the oxidation step by sodium periodate, a suitable oxidant to convert cis-glycol groups into reactive aldehyde groups. The hydrazine coupling leads to a mass increment of 26 Da for each reducing carbohydrate, broadening the molecular region for each isobaric crocins. All derivatized crocins were submitted to MS/MS experiments to clarify the structures. The MS/MS experiments of non-reductive Hydrazine tagged crocins validated several structures, confirming the presence of crocins as mixtures of different isomers. In particular, it was confirmed the presence in the natural matrix of **C8** as Hex6-Crocetin-Hex2, **C7** as Hex6-Crocetin-Hex1, **C6** as Hex6-Crocetin and Hex4-Crocetin-Hex2, **C5** as Hex3-Crocetin-Hex2 and Hex4-Crocetin-Hex1, **C4** as Hex2-Crocetin-Hex2, **C3** as Hex3-Crocetin and Hex2-Crocetin-Hex1, **C2** as Hex2-Crocetin, **C1** as Hex1-Crocetin.

Conclusions

MALDI MS was fast and selective technique to characterize crocins (glycosyl esters of crocetin) from a saline extract of saffron. The combination of mass spectrometry with a derivatization strategy able to improve ions desorption and fragmentations in MS/MS conditions, led to obtain a detailed description of the crocins structures, providing a clearer understanding of these natural molecules.

References

[1] Aiello D, Siciliano C, Mazzotti F, Di Donna L, Athanassopoulos C M, Napoli A 2020 *Food Chemistry* **307**, 125527.
[2] Risoluti R, Piazzese D, Napoli A, Materazzi S 2016 *Journal of Analytical and Applied Pyrolysis* **117**, 82-87.
[3] Aiello D, Siciliano C, Mazzotti F, Di Donna L, Athanassopoulos C M, Napoli A 2018 *RSC Advances* **8 (63)** 36104-36113.
[4] Materazzi S, Napoli A, Risoluti R, Finamore J, D'Arienzo S 2014 *International Journal of Mass Spectrometry* **365-366** 372-376.
[5] Vuono D, Guzzo M, De Luca P, Nagy J B 2014 *Journal of Thermal Analysis and Calorimetry* **116** 169–182.
[6] De Luca P, De Luca P, Candamano S, Macario A, Crea F, Nagy J B 2018 *Buildings* **8** 122.
[7] Jahouh F, Saksena R, Aiello D, Napoli A, Sindona G, Kováč P, Banoub J H 2010 *Journal of Mass Spectrometry* **45 (10)** 1148-1159.
[8] Sergi M, Gentili A, Perret D, Marchese S, Materazzi S, Curini R 2007 *Chromatographia* **65** 757-761.
[9] Gentili A, Caretti F, D'Ascenzo G, Mainero Rocca L, Marchese S, Materazzi S, Perret D 2007 *Chromatographia* **66** 669-676.
[10] Materazzi S, Risoluti R 2014 *Applied Spectroscopy Reviews* **49** 635-665.
[11] Sabbieti M G, Agas D, Santoni G, Materazzi S, Menghi G, Marchetti L 2009 *Journal of Cellular Biochemistry* **107** 316-327.
[12] Risoluti R, Gullifa G, Battistini A, Materazzi S 2019 *Analytical Chemistry* **91 (10)**, 6435-6439.
[13] De Angelis Curtis S, Kurdziel K, Materazzi S, Vecchio S 2008 *Journal of Thermal Analysis and Calorimetry* **92** 109-114.
[14] De Luca P, Bernaudo I, Elliani R, Tagarelli A, Nagy J B, Macario A 2018 *Materials* **11** 2316.
[15] Lico D, Vuono D, Siciliano C, Nagy J B, De Luca P 2019 *Journal of Environmental Management* **237** 636–643.
[16] Materazzi S, De Angelis Curtis S, Vecchio Ciprioti S, Risoluti R, Finamore J 2014a *Journal of Thermal Analysis and Calorimetry* **116** 93–98.
[17] Materazzi S, Peluso G, Ripani L, Risoluti R 2017 *Microchemical Journal* **134** 277-283.
[18] Materazzi S, Finamore J, Risoluti R, Napoli A 2014 *Microchemical Journal* **115** 27-31.
[19] Catauro M, Naviglio D, Risoluti R, Vecchio Ciprioti S 2018 *Journal of Thermal Analysis and Calorimetry* **133** 1085–1092.

[20] Risoluti R, Gullifa G, Fabiano M A, Sorrentino F, Caprari P, Materazzi S 2018c *Journal of Thermal Analysis and Calorimetry* **134** 1299–1306.
[21] De Luca P, Roberto B, Vuono D, Siciliano C, Nagy J B 2018 *IOP Conference Series: Materials Science and Engineering* **374** 012071.
[22] De Luca P, Mastroianni C, Nagy J B 2018 *IOP Conference Series: Materials Science and Engineering***374 (1)** 012003.
[23] Risoluti R, Gregori A, Schiavone S, Materazzi S 2018 *Analytical Chemistry* **90** 4288-4292.
[24] Materazzi S, Foti C, Crea F, Risoluti R, Finamore J 2014b *Thermochimica Acta* **580** 7–12.
[25] Kurdziel K, Glowiak T, Materazzi S, Jezierska J 2003 *Polyhedron* **22 (23)** 3123-3128.
[26] Risoluti R, Materazzi S, Sorrentino F, Bozzi C, Caprari P 2018a *Talanta* **183** 216–222.
[27] Catauro M, Tranquillo E, Risoluti R, Ciprioti S V 2018 *Polymers Open Access* **10 (6)** 682.
[28] Materazzi S, Risoluti R, Napoli A 2015 *Thermochimica Acta* **606** 90-94.
[29] Aiello D, Cardiano P, Cigala R M, Gans P, Giacobello F, Giuffrè O, Napoli A, Aiello D, Cardiano P, Cigala R M, Gans P, Giacobello F, Giuffrè O, Napoli A, Sammartano S 2017 *Journal of Chemical & Engineering Data* **62 (11)** 3981-3990.
[30] Materazzi S, Risoluti R, Pinci S, Romolo F S 2017a *Talanta* **174** 673-678.
[31] Materazzi S, Gregori A, Ripani L, Apriceno A, Risoluti R 2017 *Talanta* **166** 328-335.
[32] Vecchio S, Materazzi S, Wo L W, De Angelis Curtis S 2013 *Thermochimica Acta* **568** 31-37.
[33] Romolo F S, Ferri E, Mirasoli M, D' Elia M, Ripani L, Peluso G, Risoluti R, Maiolini E, Girotti S 2015 *Forensic Science International* **246** 25-30.
[34] Cardiano P, Falcone G, Foti C, Giuffrè O, Napoli A 2013 *Journal of Inorganic Biochemistry* **129** 84-93.
[35] Papadopoulos C, Cristvo B, Ferenc W, Hatzidimitriou A, Vecchio Ciprioti S, Risoluti R, Lalia-Kantouri M 2016 *Journal of Thermal Analysis and Calorimetry* **123** 717–729.
[36] Crea F, Falcone G, Foti C, Giuffrè O, Materazzi S 2014 *New Journal of Chemistry* **38** 3973-3983.
[37] Navarra M A, Materazzi S, Panero S, Scrosati B 2003 *Journal of the Electrochemical Society* **150 (11)** A1528-A1532.
[38] Bretti C, Crea F, De Stefano C, Foti C, Materazzi S, Vianelli G 2013 *Chemical Engineering Journal* **58** 2835-2847.
[39] Risoluti R, Caprari P, Gullifa G, Massimi S, Sorrentino F, Buiarelli F, Materazzi S 2019 *Microchemical Journal* **146** 374–380.
[40] Aiello D, Furia E, Siciliano C, Bongiorno D, Napoli A 2018 *Journal of Molecular Liquids* **269** 387-397.
[41] Napoli A, Athanassopoulos C M, Moschidis P, Aiello D, Di Donna L, Mazzotti F, Sindona G 2010 *Analytical Chemistry* **82 (13)** 5552-5560.
[42] Aiello D, Casadonte F, Terracciano R, Damiano R, Savino R, Sindona G, Napoli A 2016 *Oncoscience* **3 (7-8)** 220-241.
[43] Aiello D, Giambona A, Leto F, Passarello C, Damiani G, Maggio A, Siciliano C, Napoli A 2018 *Scientific Reports* **8 (1)** 10973.
[44] Imbrogno S, Aiello D, Filice M, Leo S, Mazza R, Cerra M C, Napoli A 2019 *Scientific Report* **9 (1)** 18953.
[45] Aiello D, Materazzi S, Risoluti R, Thangavel H, Di Donna L, Mazzotti F, Casadonte F, Siciliano C, Sindona G, Napoli A 2015 *Molecular BioSystems* **11 (8)** 2373-2382.
[46] Aiello D, Siciliano C, Mazzotti F, Di Donna L, Risoluti R, Napoli A 2020 *Molecules* **25 (7)** 1485.
[47] Napoli A, Aiello D, Aiello G, Cappello M S, Di Donna L, Mazzotti F, Materazzi S, Fiorillo M, Sindona G 2014 *Journal of Proteome Research* **13 (6)** 2856-2866.
[48] Salvatore L, Gallo N, Aiello D, Lunetti P, Barca A, Blasi L, Madaghiele M, Bettini S, Giancane G, Hasan M, Borovkov V, Natali M L, Campa L, Valli L, Capobianco L, Napoli A, Sannino A 2020 *International Journal of Biological Macromolecules***154** 291-306.
[49] Napoli A, Aiello D, Di Donna L, Sajjad A, Perri E, Sindona G 2006 *Analytical Chemistry* **78 (10)** 3434-3443.
[50] Temperini A, Aiello D, Mazzotti F, Athanassopoulos C M, De Luca P, Siciliano C 2020 *Molecules* **25 (6)** 25061313.

IOP Conf. Series: Materials Science and Engineering 1048 (2021) 012008 doi:10.1088/1757-899X/1048/1/012008

Latest developments in adhesive anchoring technology

P Mahrenholtz[1] **CW Hsieh**[2] **K Yu**[3]

[1] Senior Engineering Manager, Stanley Black & Decker, Dewalt Anchors, Frankfurt, Germany
[2] Professor, Department of Civil Engineering, National Pingtung Universty of Science and Technology, Pingtung, Taiwan
[3] Sales & Engineering Manager, Stanley Black & Decker, Dewalt Anchors, Taipei, Taiwan

E-mail: philipp.mahrenholtz@sbdinc.com

Abstract. Adhesive anchoring is used to connect steel and concrete elements and to connect new to existing reinforced concrete structures. For this, threaded rods or reinforcing bars are bonded to the hardened concrete by adhesives. Internationally recognized qualification guidelines and design codes form the regulative basis for all approved products and ensures a high level of safety. This paper introduces the context of adhesive anchoring systems. An overview of relevant international regulations is provided, and the difference between post-installed anchor and post-installed rebar applications is explained. The latest developments in adhesive anchoring in regard to qualification and design is presented. The contribution aims at bringing the adhesive anchoring technology to the attention of a broader audience of the engineering and research community.

1. Introduction

Post-installed adhesive anchoring systems consist of a steel element and an adhesive to bond the steel element to the concrete base material (Fig. 1a). The steel element is in general a threaded rod or a reinforcing bar (Fig. 1b). Post-installed adhesive anchors are used to connect structural and nonstructural elements to the existing concrete structure. Post-installed rebars (PIRs) may also be used for connecting new concrete to existing concrete, allowing economic construction methods (Fig. 1c).

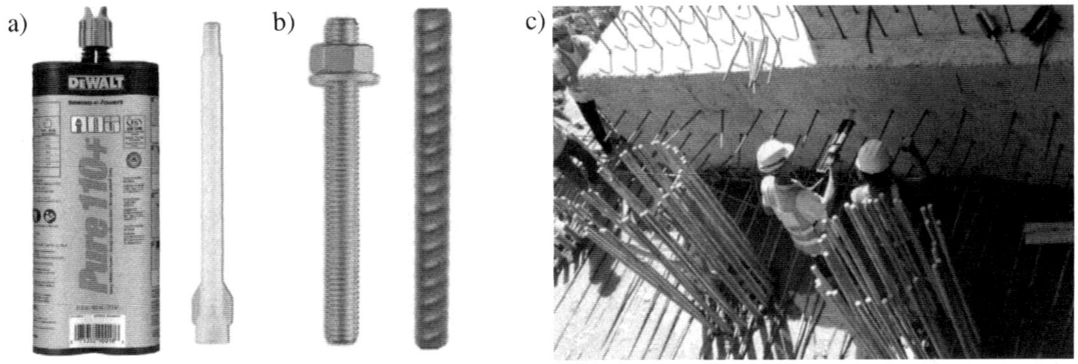

Figure. 1 a) Cartridge and mixing nozzle; b) Threaded rod and reinforcing bar; c) Starter bars installed with Dewalt PE1000+ adhesive at the M32 Highway improvement project, Taoyuan county, Taiwan

The adhesives used for anchoring in concrete are two-component systems. Due to their superior performance and slow-setting properties required in warm climates, many adhesive systems are epoxy type consisting of resin and hardener. Further adhesive systems base on epoxyacrylate, vinylester, polyester, or hybrids with some cement content; all of which fast-setting. First applications of adhesive in construction reach back half a century when bulk ware was used to mix the components on site. The commercial breakthrough of adhesive anchoring was finally experienced when cartridges allowed the injection of the exact adhesive quantity required directly into the hole drilled for installation. For this, the chemical components are filled in the compartments of the cartridges and balanced when they are dispensed through the mixing nozzle. Further information also regarding required accessories can be found in [1].

2. Background

During past fifteen years, substantial research and development have been made to increase the fields of applications and the safety of anchors and PIRs. Frequently updated qualification and design regulations reflect the latest state-of-the-art in the US and Europe. Third party testing makes approved anchoring systems very trustworthy.

2.1. Consistent system of qualification, approval and design

Third party approval documents provide design data for the conditions the anchor has been qualified for. Approved products guarantee a high level of safety due to the closed system of qualification, approval and design (Fig. 2). The qualification and design requirements in the US and Europe are largely harmonized [2]. Worldwide, designers and specifiers show an increased awareness for approved products and their benefits in view of safety and performance.

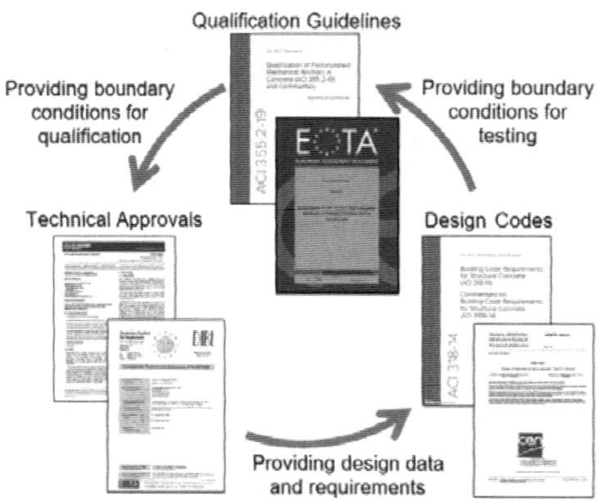

Figure. 2 Safety by qualification guidelines, technical approvals, and design guidelines (after [3])

2.2. Adhesive anchors and post-installed rebars

Engineers and designers are sometimes confused by the difference of anchor applications and PIR applications. While threaded rods are always designed as adhesive anchors, reinforcing bars can either be designed as adhesive anchors or as PIRs [4]. Adhesive anchors are designed according to anchor design rules. They typically require large edge distances due to the high bond strength they are designed for. PIRs are designed similar to cast-in reinforcing bars for development length. Since structural concrete design codes limit the maximum bond strength to relatively low values when developing the rebar, PIRs typically require deeper embedments. Because of the different underlying design assumptions, the qualification programs of adhesive anchoring systems differ for PIR and anchor applications. An overview of the relevant regulations is provided in Table 1.

Table 1. Regulations for qualification and design of PIRs in Europe and the USA

	Europe		**USA**	
	Anchor application	PIR application	Anchor application	PIR application
Qualification guideline	EAD 330499 [5]	EAD 330087 [6]	ACI 355.4 [7] Table 3.2	ACI 355.4 [7] Table 3.8
Technical approval	ETA		ESR (ICC-ES)	
Design code	EN 1992-4 [8]	EN 1992-1-1 [9]	ACI 318 [10] Chapter 17	ACI 318 [10] Chapter 25

3. Latest developments

Technical approvals of construction products like the ETA and ESR (ICC-ES) certify the performance data for the intended use of the product and an assumed service life of 50 years. For adhesive anchoring systems, performance data are primarily load capacities for the approved size range (e.g. M8 to M30 and 3/8" to 1-1/4" threaded rods, and ∅8 to ∅32 and #3 to #10 reinforcing bars), minimum and maximum embedment, as well as required edge distance and spacing. The minimum test program for adhesive anchoring systems already requires many different tests ([5], [6], [7]), including i.a. installation tests, pullout tests and creep tests at various temperatures, freeze-thaw tests, repeated load tests, tests for sensitivity to reduced cleaning effort, corrosion tests, and tests for sensitivity to sulphurous atmosphere and high alkalinity. Further tests allow the qualification for additional design applications.

3.1. 100 year design

While structural design codes generally assume a sevice life of 50 years, civil structures like bridges are sometimes designed for 100 years. For this kind of application, adhesive anchors can now be qualified for 100 year design according to the EAD 332077 [11]. For this, extra third party tests are required in addition to the basic qualification program stipulated in EAD 330499 [5]. To represent the doubled service life, critical tests are conducted with doubled test parameter. These are namely: (i) Repeated load tests where the anchor is subjected to 200,000 load cycles instead of 100,000 load cycles. A description of this test type can be found in [2]. (ii) Crack movement tests where the crack, the tested anchor is installed in, is cyclically opened and closed 2000 times instead of 1000 times. A description of this test type can be found in [12]. (iii) Sustained load tests (Fig. 3) running 6 months instead of 3 months.

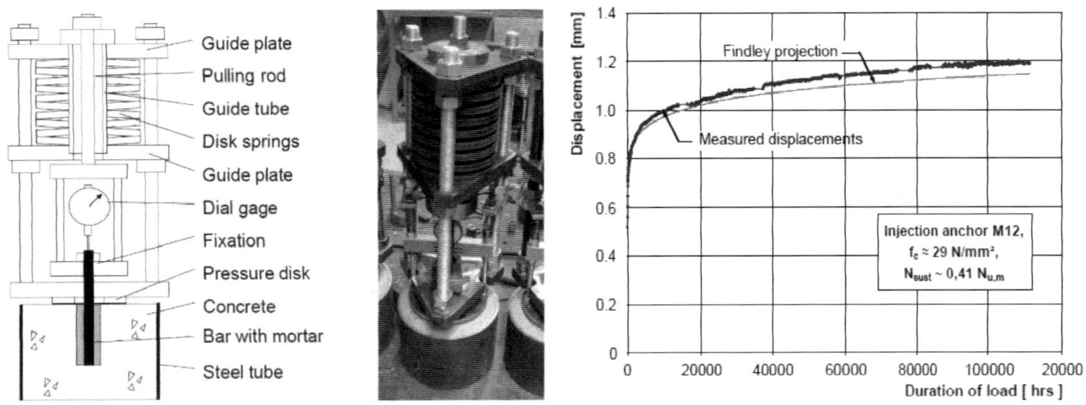

Figure. 3 a) Sustained load test setup [13]; b) Example creep curve (measured displacement over time) and Findley projection for an example anchor [14]

Sustained load tests are critical for the evaluation of the load and creep performance of adhesive anchors, and for the determination of the characteristic bond strength todesign with. They are conducted in normal ambient temperature as well as in maximum long term temperature. For this, the installed anchor is loaded over the required test duration (Fig. 3a), and the measured displacement is extrapolated using the Findley projection (Fig. 3b) and compared against reference values.

3.2. Seismic C2 qualification

Seismic loads are one of the most severe but least predictable actions structures and attached components may be subjected to. Seismic qualification of anchors has been covered in the US for long time but only recently introduced in Europe as a 2-tiered qualification requirement within the EAD 330499 guideline [15]. The qualification for a higher seismicity or higher risk category (akin to building importance) is denominated C2. The superior performance of C2 anchors is not expressed by better design parameters, such as increased design strength and decreased design displacement, but rather verified by a more rigorous test regime accounting for more adverse test conditions. To qualify for C2, anchors have to pass successfully simulated seismic tests with extreme cyclic anchor load and crack width regimes (Fig. 4a). To further increase safety, the introduction of C2 qualification requirements for ICC-ES approvals is currently debated in the US [16].

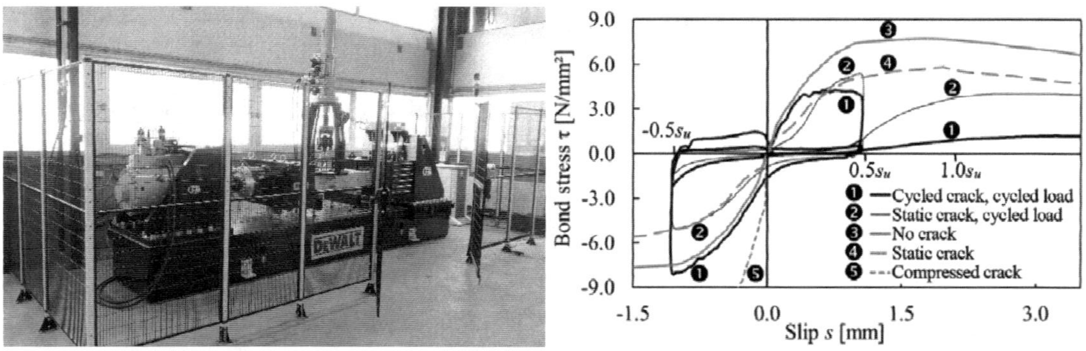

Figure. 4 a) Seismic test rig [16]; b) Example bond-slip curve on reinforcing bar [4]

The other way around, seismic qualification of PIRs has been feasible for ICC-ES approvals according to ACI 355.4 for quite some time, but for ETA only with the publication of the the EAD 331522 [17]. For this, PIRs are subjected to seismic loading and their performance compared to that of cast-in reinfording bars. (Fig. 4b). Further, the minimum concrete cover can be determined by splitting tests.

3.3. Fire rating

Fire catastrophies are frequently recurring events with fatal consequences in particular for high-rise buildings. Fire ratings of critical building components are important for passive fire safety. The fire performance is evaluated for specific fire durations, typically ranging from 30 to 120 minutes (F30 to F120). In case of adhesive anchors, the duration till anchor failure is determined by fire tests according to EOTA TR 020 [18]. For this, several anchors are installed overhead in a concrete slab and loaded by dead weights while burned fuel heats the furnace according to the standard temperature curve (Fig. 5a). The fire design strength is provided in a fire report. For fire rating of PIRs according to EAD 330087 [6], reinforcing bars are installed in a concrete cylinder that is heated by an electrical furnace, and hydraulically loaded to failure (Fig. 5b). Test series at variable temperatures result in a bond reduction curve published in the ETA. This curve is then taken as input parameter for FE simulations of specific design situations and fire durations. The determined fire design strength is again provided in a fire report.

Figure. 5 a) Fire test on anchors (example picture taken from fire report on Dewalt Pure110 adhesive [19]); b) Fire test on PIRs (example picture taken from fire report on Dewalt PE1000 adhesive [20])

Fire qualification of bonded anchors and PIRs is complex [21] and subject of ongoing research. Fire performance of adhiesves are progressively improved and high performing adhesives are now qualified for up to 240 minutes fire (F240) in PIR applications. Due to missing regulations in the US, US design professionals fall back on the European fire reports.

4. Conclusion

Adhesive systems allow state-of-the-art anchoring in concrete with unprecedented flexibility. The consistent system of qualification guidelines, technical approvals, and design codes ensures a safe connection between steel and concrete (anchor), and concrete and concrete (PIR). US and European regulations are most advanced and to a great extent harmonized. Because of their extreme testing requirements US and European regulations are internationally recognized; and third party testing makes products approved by an ETA or ESR very trustworthy (Fig. 6).

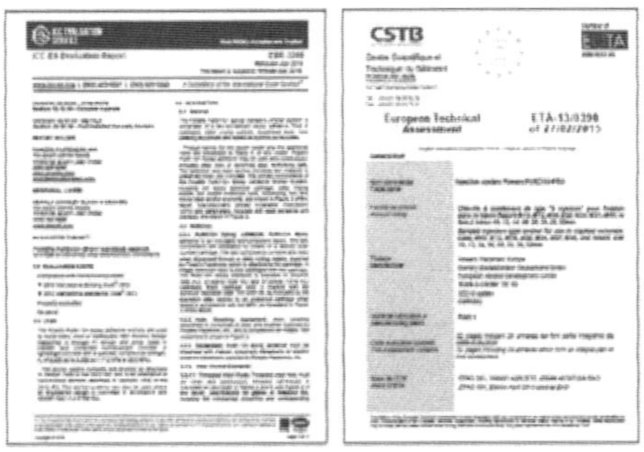

Figure. 6 Technical approval documents: Front pages of a) ETA and b) ESR (ICC-ES)

Products and regulations are permanently advancing. The latest product generation are high performing epoxies that are approved beyond regular applications of anchors and as PIRs. Recent code developments allow to expand the design service life from 50 to 100 years based on extended testing. New seismic requirements have been introduced to qualify anchors for the performance category C2, and PIRs for seismic applications. Fire tests on anchors and PIRs provide ratings for up to 240 min fire durations. – Approved adhesive anchoring systems are surely tested to extremes.

Opinions, conclusions, and recommendations expressed in this paper are those of the authors only and do not necessarily reflect those of the authors' affiliations or other sponsoring agencies.

References

[1] Mahrenholtz P, Olsen J. Background, regulations and latest developments of post-installed adhesive anchors and rebars. Proceedings of the 3rd International Symposium ConSC2017 (Connections between Steel and Concrete), Stuttgart 2017.
[2] Mahrenholtz P, Wood R. Design of post-installed and cast-in-place anchors according to the new EN 1992-4 and ACI 318-19. Structural Concrete, Journal of the fib, in press 2020.
[3] Mahrenholtz P. Experimental performance and recommendations for qualification of post-installed anchors for seismic applications. Dissertation, University of Stuttgart 2012.
[4] Mahrenholtz C, Eligehausen R, Reinhardt H-W. Qualification and design of cast-in-place and post-installed reinforcing bar anchorages. ACI Structural Journal, March-April 2020, Vol. 117, No. 2, 3-16 2020.
[5] EAD 330499-01-0601. Bonded fasteners for use in concrete. European Assessment Document, Pending for citation in OJEU 2020.
[6] EAD 330087-00-0601. Systems for post-installed rebar connections with mortar. European Assessment Document, OJEU 2018/C 370/05, EOTA 2018.
[7] ACI 355.4. Qualification of post-installed mechanical anchors in concrete (ACI 355.2-19) and commentary, American Concrete Institute (ACI), Farmington Hills, Michigan 2019.
[8] EN 1992-4. Eurocode 2: Design of concrete structures – Part 4: Design of fastenings for use in concrete. European Committee for Standardization (CEN); EN 1992-4 2018.
[9] EN 1992-1-1. Eurocode 2: Design of concrete structures – Part 1-1: General rules, and rules for buildings. European Committee for Standardization (CEN); EN 1992-1-1:2004 2004.
[10]ACI 318. Building code requirements for structural concrete (ACI 318-19) and commentary (ACI 318R-19), American Concrete Institute, Farmington Hills, Michigan 2019.
[11]EAD 332077-00-0601. Bonded fasteners for use in concrete with service life of 100 years. European Assessment Document, Pending for citation in OJEU 2019.
[12]Mahrenholtz C, Sharma A. Testing of anchors and reinforcing bars in concrete under cyclic crack movements. ASTM Journal of Testing and Evaluation, Vol. 45, Issue 4, July 2017, 1326-1337 2017.
[13]Mahrenholtz C, Eligehausen R, Hofmann J, Fuchs W. Post-installed bars under low installation temperatures. Proceedings of the 3rd International Conference on Concrete Repair, Rehabilitation and Retrofitting (ICCRRR), Cape Town, 1195–1201 2012.
[14]Eligehausen R, Blochwitz R, Fuchs W. Behavior and design of adhesive anchors under sustained load. ACI Special Publication 283, 9.1-9.13 2012.
[15]Mahrenholtz P, Wood R, Eligehausen R, Hutchinson T, Hoehler M. Development and validation of European guidelines for seismic qualification of post-installed anchors. Engineering Structures 148, 497-508 2017.
[16]Mahrenholtz P, Wood R. European seismic performance categories C1 and C2 for post-installed anchors. ACI Structural Journal, in review 2020.
[17]EAD 331522-00-0601. Post-installed rebar with mortar under seismic action. European Assessment Document, Pending for citation in OJEU 2018.
[18]EOTA TR 020. Evaluation of anchorages in concrete concerning resistance to fire. Technical Report 020 of the European Organization of Technical Approvals (EOTA), Brussels 2006.
[19]MFPA Leipzig. Report No. 3.2/13-055 of fire tests on Pure110 adhesive 2013.
[20]CSTB Paris. Report No. 26048096 of fire tests on Pure150/PE1000 adhesive 2014.
[21]Mahrenholtz P, Sharma A. Fire rating of post-installed anchors and rebars. Connections in concrete, special issue of CivilEng Journal by MDPI, in preparation 2020.

IOP Conf. Series: Materials Science and Engineering 1048 (2021) 012009 doi:10.1088/1757-899X/1048/1/012009

Identification and isolation of homoserine lactones (HSLs) produced by Pseudomonas aeruginosa and the effects on Legionella pneumophila growth

S. Materazzi[1]*, M. De Giusti[2], L. Marinelli[2]

[1]Department of Chemistry "Sapienza" University of Rome
[2]Department of Public Health and Infectious Disease, Sapienza" University of Rome

E-mail: stefano.materazzi@uniroma1.it

Abstract. Risk for infections from Legionella pneumophila represents a critical issue for immunocompromised individuals, especially when this species is present within the biofilm of the water distribution systems of hospitals or other health facilities where the complex ancient water network creates environmental conditions that are suitable for Legionella persistence and multiplication. The shock hyperchlorination and subsequent systemic continuous chlorination is a disinfection treatment reported in the Italian Guideline for prevention and control of nosocomial legionellosis as an effective procedure to reduce L. pneumophila colonization of hospital water sources. Although this disinfection strategy has been proven to be effective in the short-term, eradication of Legionella from water networks is usually not achieved. The objective of this study is the identification and purification of P. aeruginosa bacteriocins, such as homoserine lactones (HSLs), by the development of an experimental method by High performance-liquid chromatography (HPLC), coupled to diode array detection (DAD) and the evaluation of the effectiveness of these molecules in the managing of the persistence of L. pneumophila in experimental cocultures.

1. Introduction

Legionella pneumophila is an ubiquitous microorganism present in both natural and artificial water systems that may cause a potential life threatening form of pneumonia called Legionnaires disease, especially when colonizing the water distribution systems of buildings of hospitals or other health facilities, the risk for infections for immunocompromised individuals increases greatly [1-6]. In such artificial aquatic systems, the biofilm covering the interior of pipelines and, more in general, the various plumbing elements represent relatively nutrient-rich spots where Legionella (and other bacteria) can attach and multiply [7-13]. Biofilm is a diversified and complex matrix attached to any natural or artificial surface. It is composed by heterogeneous aggregates of organisms and is primarily composed by bacteria, fungi, protozoa, and algal populations and associated extracellular polymeric substances [14-22]. The bacterial cells embedded in such matrix are connected by interstitial water channels, allowing transport (and probably exchanges) of nutrients, oxygen, genes and even antimicrobial agents between micro-colonies. An interspecific mechanism of signalling between microbial cells (e.g. quorum sensing) might also be involved in cross-regulating the growth of different species [23-31]. As a result, many engineered water systems such as air conditioning cooling towers, water boilers, whirlpools and spas, drinking water distribution networks, shower heads, and dental-unit water lines provide an environment conducive to the growth and multiplication of the Legionella spp. The survival of microorganisms in man-made systems results from the (interactive) effects of several factors which depend themselves from the structure and management of the water network [32-39]. Those factors include temperature, pipe material, nutrient levels, frequencies, type and concentration of disinfectants, water velocity and hydraulic conditions. Since those factors are more or less variable in time and space, the distribution of biofilm and its taxa composition is not uniform too making difficult to predict the presence of L. pneumophila in a given water network [40-43]. Several empirical studies have examined the simultaneous effects of chemical elements, biofilm biomass (often quantified as heterotrophic plate counts, HPC, at 22° and / or 37°) and other factors on L. pneumophila presence [44-52]. Among the micronutrients, lower levels of certain metals such as iron, zinc, and potassium enhance growth of L. pneumophila. Several papers have shown that drinking water bacteria often possess the ability to inhibit Legionella growth. More recently, it has been observed that monospecies biofilm of P. aeruginosa is not conductive to L. pneumophila persistence under dynamic flow conditions possibly because of inhibitory substances secreted by the former (e.g. lactones) [53-55].

2. Experimental

Materials and methods

Reference standard solution of 3-oxo-C10 homoserine lactone (3 -oxo-C10HSL), 3-oxo-C12 homoserine lactone (3 -oxo-C12HSL) and 3-oxo-C14 homoserine lactone (3 -oxo-C14HSL) (purity≥98%) were purchased from Sigma Aldrich. Acetonitrile, acetic acid, methanol and ultrapure water (HPLC-grade) were used for the chromatographic separation while dichloromethane and ethyl acetate (purity≥98%) were used for the extraction of HSL prior to HPLC injection. The analyses were performed by a Shimadzu HPLC equipped with a diode array (DAD), a dual solvent pump and a Rheodyne injection valve. Separation and detection of HSLs was achieved with a C18 column (5 µm, 250 mm x 4.60 mm) maintained at 40°C. The mobile phase was acetonitrile-water 75-25 %vv) at a flow rate of 1 ml/min.

Study design and bacterial strains

To test the effect of the HSL on the Legionella growth, a number of experiments were performed in order to evaluate the viability of the cells by the enumeration of the viable cells on selective agar at different periods and by the optical density. The bacteria used in co-cultures were isolated from tap water of the hospital according to the standard method (UNI EN ISO 6222:2001). The Legionella strain used in this study was the same serogroup isolated from the tap water of the hospital where the other bacteria will be also obtained [56-62].

3. Results

Cultures of L. pneumophila and P. aeruginosa were inoculated in specific media and allowed growing for 16-18 hours at 37 ° C under stirring at 200 RPM. The supernatants of P. Aeruginosa were simultaneously centrifuged, filtered and subjected to extraction for the determination of homoserine compounds. The extracts of the supernatants of Pseudomonas were analyzed by high performance liquid chromatography (HPLC-DAD) and the calibration curves for the quantification was constructed for each analyte. Results are reported in Figure 1, where the resulting chromatogram shows a suitable separation of the HSLs that would permit the further collection of the different fractions for the inoculation in Legionella co-coltures. Satisfactory calibration curves were achieved for all the investigated molecules with a correlation coefficient not lower than 0.999. The HSLs identified by the optimized HPLC method were inoculated in the coltures of L. pneumophila in order to study the effectiveness in the reduction of the persistency of Legionella. The optical density was measured at 600 nm and the results were correlated to the number of viable cells determined by the estimation of the colony forming units (CFU). The Figure 2a reports the evaluation of the optical density for L. pneumophila (red) considered as reference and compared with the optical density obtained for L. pneumophila over a period of 24 h, inoculated with the supernatant of P. aeruginosa (blue) and with P. fluorescens (green). The inoculation with P. aeruginosa exhibited the highest inhibition effect on the persistency of L. pneumophila and the enumeration of the viable cells reported in Figure 2b, confirmed the lowest values of cells growth when the supernatant of P. aeruginosa was used in the co-coltures. As a consequence, this effect may be related to the presence of the 3-oxo-C10 homoserine lactone), 3-oxo-C12 homoserine lactone and 3-oxo-C14 homoserine lactone isolated and characterized by the HPLC method. In addition, the effect of the HSL compounds on the Legionella growth was evaluated over a period of 6 days and the results of the optical density and the enumeration of the living cells are reported in Figure 2b. Coltures of Legionella inoculates with P. aeruginosa provided for the highest bacteriocin effect, exhibiting the lowest values of absorbance and colony forming unit of Legionella compared to the untreated one.

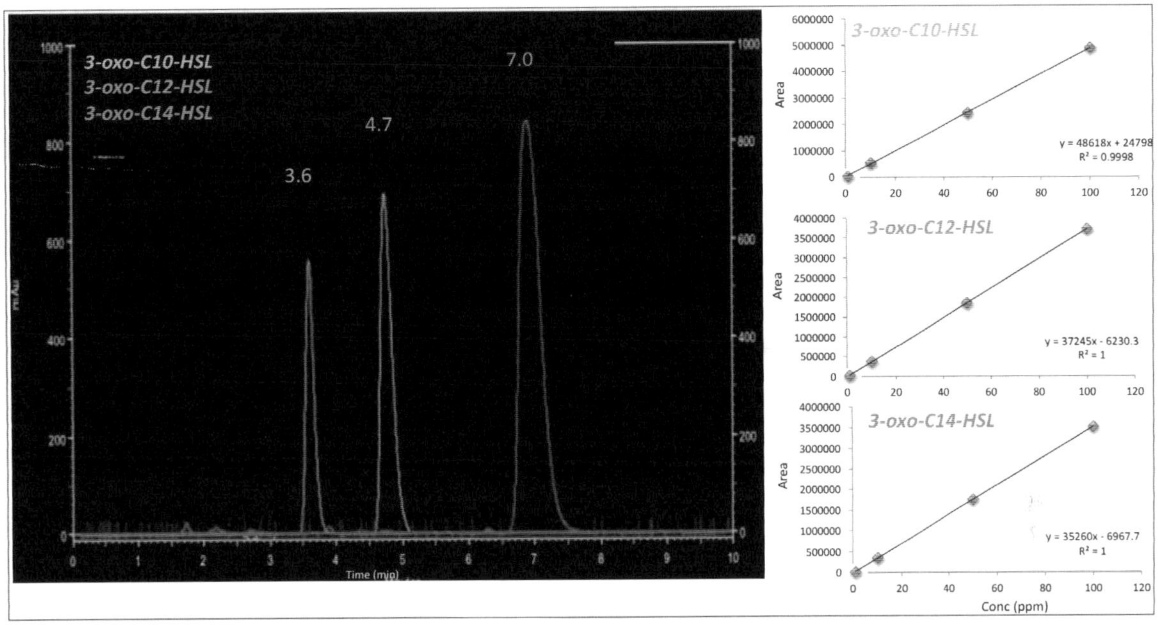

Figure 1. a) Chromatographic separation of 3-oxo-HSL by HPLC-UV; b) Calibration curves obtained for quantitation of each 3-oxo-HSL

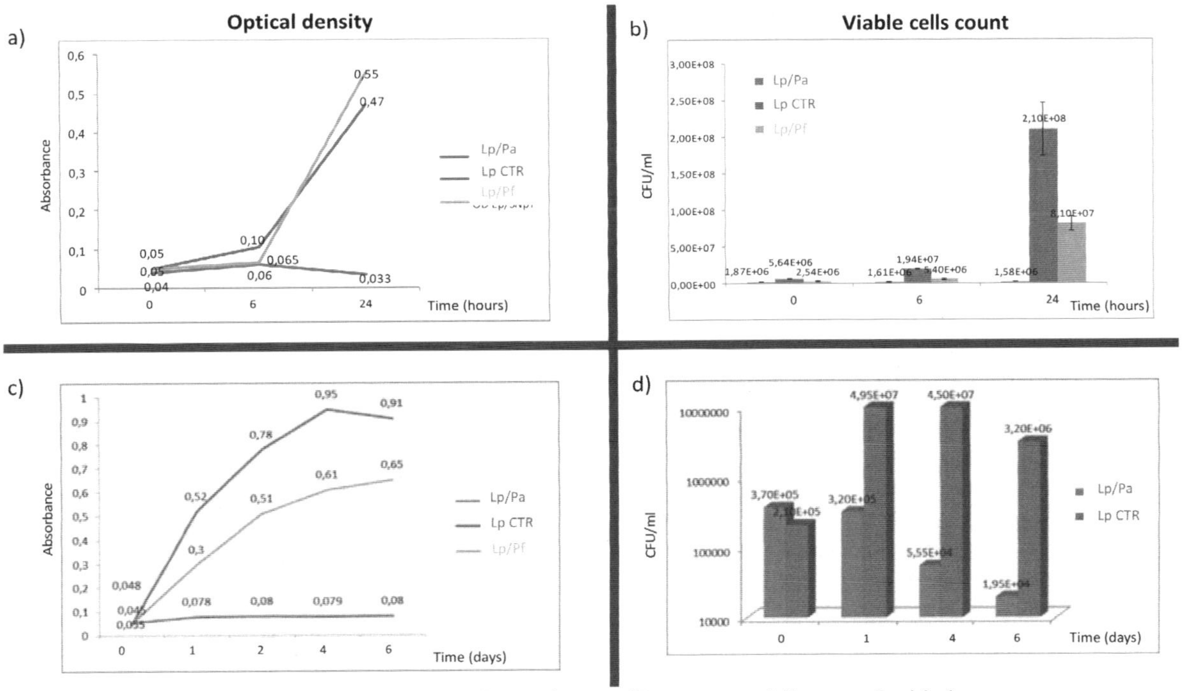

Figure. 2 Monitoring of the optical density of co-cultures of L. pneumophila treated with the supernatant of P. aeruginosa (blue) and P. fluorescens (green) over a period of 24 h (a) and 6 days (c); Monitoring

of the CFU of co-cultures of L. pneumophila treated with the supernatant of P. aeruginosa (blue) over a period of 24 h (b) and 6 days (d)

Conclusions

In this work, a novel HPLC-UV method was optimized for the isolation and characterization of homoserine lactones HSLs in P. aeruginosa supernatant. The separated and collected fractions of 3-oxo-C10 homoserine lactone (3 -oxo-C10HSL), 3-oxo-C12 homoserine lactone (3 -oxo-C12HSL) and 3-oxo-C14 homoserine lactone (3 -oxo-C14HSL) were further used to inoculate cultures of L. pneumophila with the aim of evaluating the inhibition effect against Legionella growth. Results demonstrated that the supernatant of P. aeruginosa produces HSLs that are responsible of the inhibition of the persistence of Legionella. In fact, the evaluation of the optical density and the enumeration of the colony forming units (CFU) suggested that the treatment of L. pneumophila biofilm with HSLs may reduce the persistency of Legionella.

References

[1]. Khaledi A, Esmaeili SA, Vazini H, Karami P, Bahrami A, Sahebkar A 2019 Evaluation of the prevalence of Legionella pneumophila in Iranian clinical samples: A systematic review and meta-analysis. *Microbial Pathogenesis* **129**, 93-98

[2]. Aiello D, Materazzi S, Risoluti R, Thangavel H, Di Donna L, Mazzotti F, Casadonte F, Siciliano C, Sindona G, Napoli A 2015 *Molecular BioSystems* **11 (8)**, 2373-2382.

[3]. Risoluti R, Piazzese D, Napoli A, Materazzi S 2016 *J Anal ApplPyrol* **117**, 82-87.

[4]. Cardiano P, Falcone G, Foti C, Giuffrè O, Napoli A 2013 *J Inorg Biochem* **129**, 84-93.

[5]. Reddy C E, Albanito L, De Marco P, Aiello D, Maggiolini M, Napoli A, Musti A M 2013 *Cell Death and Dis* **4 (10)**, e852.

[6]. Materazzi S, Risoluti R, Napoli A 2015 *Thermochim. Acta* **606**, 90-94.

[7]. March GA, Gutierrez MP, Lopez I, Munoz MF, de Lejarazua RO, Simarro M, Orduna A, Bratos MA 2003–2016, 2019 *Enfermedades Infecciosas y Microbiología Clínica* **37(8),** 514-520

[8]. Mazzotti F, Benabdelkamel H, Di Donna L, Athanassopoulos C M, Napoli A, Sindona G 2012 *J Mass Spectrom* **47 (7)**, 932-939.

[9]. Cardiano P, Giuffre O, Napoli A, Sammartano S 2009 *New J Chem* **33 (11)**, 2286-2295.

[10].Jahouh F, Saksena R, Aiello D, Napoli A, Sindona G, Kováč P, Banoub J H 2010 *J Mass Spectrom.* **45 (10)**, 1148-1159.

[11].Napoli A, Athanassopoulos C M, Moschidis P, Aiello D, Di Donna L, Mazzotti F, Sindona G 2010 *Anal. Chem.* **82 (13)**, 5552-5560.

[12]. Aiello D, Cardiano P, Cigala R M, Gans P, Giacobello F, Giuffrè O, Napoli A, Sammartano S 2017 *J Chem. Eng Data* **62 (11),** 3981-3990.

[13]. Materazzi S, Finamore J, Risoluti R, Napoli A 2014 *Microchem J* **115**, 27-31.

[14]. Swart AL, Hilbi H 2020 *Legionella pneumophila.* Reference Module in Biomedical Sciences

[15].Materazzi S, Risoluti R, Pinci S, Romolo F S 2017a *Talanta* **174**, 673-678.

[16].Aiello D, Giambona A, Leto F, Passarello C, Damiani G, Maggio A, Siciliano C, Napoli A 2018 *Sci. Rep.* **8 (1)**, 10973.

[17]. Curcio R, Aiello D, Vozza A, Muto L, Martello E, Cappello A R, Capobianco L, Fiermonte G, Siciliano C, Napoli A, Dolce V 2020 *Mol. Biotechnol.* **62 (2)**, 119-131.

[18]. Beneduci A, Corrente G A, Marino T, Aiello D, Bartella L, Di Donna L, Napoli A, Russo N, Romeo I, Furia E 2019 *J. Mol. Liq.* **296**, 111805.

[19].Imbrogno S, Aiello D, Filice M, Leo S, Mazza R, Cerra M C, Napoli A 2019 *Sci. Rep.* **9 (1)**, 18953.

[20]. Chillè D, Aiello D, Grasso G I, Giuffrè O, Napoli A, Sgarlata C, Foti C, 2020 *J Environ Sci (China)* **94**, 100-110.

[21]. Salvatore L, Gallo N, Aiello D, Lunetti P, Barca A, Blasi L, Madaghiele M, Bettini S, Giancane G, Hasan M, Borovkov V, Natali M L, Campa L, Valli L, Capobianco L, Napoli A, Sannino A 2020 *Int. J. Biol. Macromol.* **154**, 291-306.

[22]. Bartella L, Mazzotti, F, Sindona G, Napoli A, Di Donna L 2020 *FOOD CHEM TOXICOL* **136**, 111110.

[23]. S. Materazzi, A. Gregori, L. Ripani, A. Apriceno, R. Risoluti. *Talanta* 2017, **166**, 328-335
[24].Aiello D, Siciliano C, Mazzotti F, Di Donna L, Athanassopoulos C M, Napoli A 2018 *RSC Adv* **8 (63)**, 36104-36113.
[25]. Mazzotti F, Di Donna L, Napoli A, Aiello D, Siciliano C, Athanassopoulos C M, Sindona G 2014 *J Mass Spectrome.* **49 (9)**, 802-810
[26]. Napoli A, Aiello D, Di Donna L, Moschidis P, Sindona G 2008 *J. Proteome Res.* **7 (7),** 2723-2732.
[27]. Crea F, Falcone G, Foti C, Giuffrè O, Materazzi S 2014 *New J. Chem.* **38**, 3973-3983.
[28]. Materazzi S, Napoli A, Risoluti R, Finamore J, D' Arienzo S 2014 *Int. J. Mass Spectrom.* **365-366**, 372-376.
[29].Bretti C, Crea F, De Stefano C, Foti C, Materazzi S, Vianelli G 2013 *Chem. Eng. J.* **58**, 2835-2847.
[30]. Agas D, Marchetti L, Menghi G, Materazzi S, Materazzi G, Capacchietti M, Hurley M M, Sabbieti M G 2008 *J. Cell. Physiol.* **214**, 145-152.
[31]. Gentili A, Caretti F, D'Ascenzo G, Mainero Rocca L, Marchese S, Materazzi S, Perret D 2007 *Chromatographia* **66**, 669-676.
[32].Arslan EO, Kimiran AA 2018 An investigation of virulence factors of Legionellapneumophila environmental isolates *Journal of Brazilian Microbiology* **49**, 189-199
[33].Sergi M, Gentili A, Perret D, Marchese S, Materazzi S, Curini R 2007 *Chromatographia* 65, 757-761.
[34]. Aiello D, Giambona A, Leto F, Passarello C, Damiani G, Maggio A, Siciliano C, Napoli A 2018 *Sci Rep* **8 (1),** 10973
[35].Napoli A, Aiello D, Aiello G, Cappello M S, Di Donna L, Mazzotti F, Materazzi S, Fiorillo M, Sindona G 2014 *J. Proteome Res.* **13**, 2856-2866.
[36]. Vecchio S, Materazzi S, Wo L W, De Angelis Curtis S 2013 *Thermochim. Acta* 568, 31-37.
[37]. Materazzi S, De Angelis Curtis S, Vecchio Ciprioti S, Risoluti R, Finamore J 2014 *J. Therm. Anal. Calorim.* 116, 93–98.
[38]. Aiello D, Siciliano C, Mazzotti F, Donna L D, Risoluti R, Napoli A 2020 *Molecules* **25 (7),** 1485.
[39].De Angelis Curtis S, Kurdziel K, Materazzi S, Vecchio S 2008 *J. Therm. Anal. Calorim.* **92**, 109-114.
[40]. Cataldi TR, Bianco G, Abate S 2008 *J. Mass Spectrom.* **43**, 82–96
[41].Materazzi S, Peluso G, Ripani L, Risoluti R 2017 *Microchem. J.* 134, 277-283.
[42]. Risoluti R, Gregori A, Schiavone S, Materazzi S 2018 *Anal. Chem.* 90, 4288-4292.
[43]. Aiello D, Furia E, Siciliano C, Bongiorno D, Napoli A 2018 *J Mol Liq.* **269**, 387-397.
[44].Ortori CA, Halliday N, Cámara M, Williams P , Barrett D 2011 LC-MS/MS Quantitative Analysis of Quorum Sensing Signal *Molecules Bioanal Chem.* **399**, 839–850.
[45].Risoluti R, Fabiano M A, Gullifa G, Vecchio Ciprioti S, Materazzi S 2017 *Appl. Spectrosc. Rev.* 52, 39-72.
[46]. Navarra M A, Materazzi S, Panero S, Scrosati B 2003 *J. Electrochem. Soc.* **150 (11)**, A1528-A1532.
[47].Materazzi S, Foti C, Crea F, Risoluti R, Finamore J 2014 *Thermochim. Acta* **580**, 7–12.
[48]. Marchetti L, Sabbieti M G, Menghi M, Materazzi S, Hurley M M, Menghi G 2002 *Histol Histopathol* **17 (4),** 1061-1066.
[49]. Kurdziel K, Glowiak T, Materazzi S, Jezierska J 2003 *Polyhedron* **22 (23)**, 3123-3128.
[50]. Aiello D, Siciliano C, Mazzotti F, Di Donna L, Athanassopoulos C M, Napoli A 2020 *Food Chem.* **307**, 125527.
[51].Papadopoulos C, Cristvo B, Ferenc W, Hatzidimitriou A, Vecchio Ciprioti S, Risoluti R, Lalia-Kantouri M 2016 *J. Therm. Anal. Calorim.* **123**, 717–729.
[52]. Romolo F S, Ferri E, Mirasoli M, D' Elia M, Ripani L, Peluso G, Risoluti R, Maiolini E, Girotti S 2015 *Forensic Sci Int.* 246, 25-30
[53].Tateda K, Ishii Y, Horikawa M, Matsumoto T, Miyairi S, Pechere JC, Standiford TJ, Ishiguro M, Yamaguchi K 2003 The Pseudomonas aeruginosa Autoinducer N-3-Oxododecanoyl Homoserine Lactone Accelerates Apoptosis in Macrophages and Neutrophils *Infect Immun.* **71**, 5785–5793.
[54].Catauro M, Naviglio D, Risoluti R, Vecchio Ciprioti S 2018 *J. Therm. Anal. Calorimetry* **133**, 1085–1092.

[55]. Risoluti R, Materazzi S, Sorrentino F, Bozzi C, Caprari P 2018 *Talanta* **183**, 216–222.
[56]. Sharma S, Sachdeva P, Virdi JS 2003 Emerging water-borne pathogens. *Appl Microbiol Biotechnol.* **61**, 424–428
[57].Risoluti R, Gullifa G, Fabiano M A, Sorrentino F, Caprari P, Materazzi S 2018c *J. Thermal Analys. Calorim.* **134**,1299–1306.
[58]. Aiello D, Casadonte F, Terracciano R, Damiano R, Savino R, Sindona G, Napoli A 2016 *Oncoscience* **3 (7-8)**, 220-241.
[59].Risoluti R, Gullifa G, Battistini A, Materazzi S 2019 *Anal Chem* 91 (10), 6435.
[60].Materazzi S, Vecchio S, Wo LW, De Angelis Curtis S 2012 *Thermochimica Acta*, **543**, 183-187
[61].Migliorati V, Ballirano P, Gontrani L, Materazzi S, Ceccacci F, Caminiti R. 2013 *Journal of Physical Chemistry B*, **117**, 7806-7818
[62].Materazzi S, Risoluti R 2014 *Applied Spectroscopy Reviews*, **49 (8)**, 635-665.

Stability analysis of intrinsic camera calibration using probability distributions

Kun Zhan[1], Dieter Fritsch [2] and Joerg F Wagner [3]

[1] Adaptive Structures in Aerospace Engineering，University of Stuttgart，Pfaffenwaldring 31, D-70569 Stuttgart, Germany
[2] Institute for Photogrammetry， University of Stuttgart, Geschwister-Scholl-Str. 24D, D-70174 Stuttgart, Germany
[3] Adaptive Structures in Aerospace Engineering，University of Stuttgart, Pfaffenwaldring 31， D-70569 Stuttgart, Germany

[1] kun.zhan@pas.uni-stuttgart.de
[2] dieter.fritsch@ifp.uni-stuttgart.de
[3] jfw@pas.uni-stuttgart.de

Abstract. The currently most widely used calibration method in geometric computer vision is based on Zhang's method. This approach is easy to implement and delivers detailed result evaluations. One drawback, however, might be that the unexperienced user may obtain unstable and unrepeatable calibration results despite small re-projection errors. Therefore, the data must be interpreted with caution and there is definitely a need for further consideration beside re-projections to be taken for better estimations. So far, statistical inference has been used for presenting precision measures only. In our work, we extend the statistical approach quantitatively using large bundles of images and get calibrations from randomly chosen subsets. Then a Maximum Likelihood Estimation of intrinsic parameters is implemented and the statistical behaviour is analyzed. In addition, the recovered expected values of parameters are utilized as ground truth to scrutinize the single influencing factors of the imaging configuration. According to the results we found out that the image block plays a significant role for camera calibration with regard to the orientation of the imaging configuration. Including also suitable manufactured calibration boards as well as other operational issues, finally a well-designed image block provides a repeatable and reliable calibration.

1. Introduction

Camera calibration is one main focus of airborne and close-range photogrammetry. Due to image sensor and lens imperfections, camera calibration is a crucial step in many fields for metric scene measurements. The imperfections lead to deviations of the projection relationships between image and object space and must be compensated. For decades, camera systems are calibrated regarding not only the interior parameters, but also lens distortion, by using additional parameters based on physical and mathematical models [1]. It took decades to prove mathematically, that camera calibration belongs to the problems of algebraic function approximation, as given in detail by [2], [3] and [4].

In geometric computer vision a plane point field calibration with Zhang's method [5] is widely used, due to its mature algorithm. A comparative review of earlier related works can be found in [6]. In [7] the merits and drawbacks of different methods are illustrated. Test field calibration can work with 3D test-fields as given in [8], which might, however, require high manufacturing expenses. Zhang's calibration method requires at least 5 views of images of a chess board from different orientations. After the feature extraction, the projective transformation is calculated for every image. A closed-form solution is applied to derive the initial calibration results and then all parameters including additional distortion parameters will be refined by a global optimization method [5]. Although the method is easy to implement and normally can get high accuracy, users without any related background will end up with unsatisfying accuracies or with small re-projection errors and unstable calibration parameters as stated in [9]. Thus, among many factors, the set-up for image data collection is most important for camera calibration. This is confirmed by many references about image block configurations for camera calibration, as presented in [10], [11].

Our work is proposing a method to quantitatively analyze the different factors to give finally specific suggestions on camera calibration even for users without little prior knowledge. It is triggered by finding a simple but rigorous calibration method for the 3D digitization of the gyroscopic instruments collection of the University of Stuttgart – the Gyrolog project [12].

The ground truth of calibration parameters in reality is unknown and can be approximated only. However, the image configuration and feature extraction process can be assumed to have a normal distribution behavior [13]. Therefore, through a normal distribution estimation from a big amount of calibration experiments, it is possible to recover the expected values. Thus, it can be regarded as ground truth estimating the precision of individual calibration [14]. In this paper, the new calibration evaluation method proposed uses the probability distribution. Moreover, quantitative influences are also analyzed to give practical suggestions.

2. Materials and Methods

In computer vision, the calibration precision is not of high demand. Therefore, the widely-used calibration tools do not deliver strategies for image data collection, which actually will cause either unsatisfied calibration accuracies or small re-projection errors. In [15] the camera calibration wizard delivers some guidance on how to compute the next best pose for calibrations. The method is based on a subsequent optimization process to obtain the next best image for a good calibration. This method has not been widely adopted yet. Most of the methods use well-designed 3D test fields, which provide for better ray intersections, but the expensive costs may be a restriction. In our method, different factors are quantitatively experimented and analyzed and more specific suggestions are given.

2.1. Related Theory

2.1.1. Camera Calibration. For the relationships between the 2D and 3D world in Euclidean geometry, different coordinate systems are defined. At the beginning, the camera model is assumed as an ideal camera model without distortion, which complies with a direct linear transformation.

$$\begin{bmatrix} u \\ v \\ 1 \end{bmatrix} = \begin{bmatrix} 1/dX & 0 & u_0 \\ 0 & 1/dY & v_0 \\ 0 & 0 & 1 \end{bmatrix} \begin{bmatrix} f & 0 & 0 & 0 \\ 0 & f & 0 & 0 \\ 0 & 0 & 1 & 0 \end{bmatrix} \begin{bmatrix} \mathbf{R} & \mathbf{t} \\ 0 & 1 \end{bmatrix} \begin{bmatrix} X \\ Y \\ Z \\ 1 \end{bmatrix} \tag{1}$$

Here X, Y, Z are the object coordinates in the world coordinate system, $\mathbf{R}$ and $\mathbf{t} = [X_0 \quad Y_0 \quad Z_0]^T$ are representing rotations and translations, respectively, between the world coordinate system and the camera coordinate system, f stands for the focal length, dX, dY are scale factors in the image frame, (u_0, v_0) are principal point coordinates and u, v are the pixel coordinates. When extending the ideal lens characteristics with distortion, it can be written into the collinearity equations.

$$\mathrm{u} = u_0 + f\frac{r_{11}(X - X_0) + r_{12}(Y - Y_0) + r_{13}(Z - Z_0)}{+r_{13}(X - X_0) + r_{23}(Y - Y_0) + r_{33}(Z - Z_0)} + \Delta u \tag{2}$$

$$\mathrm{v} = v_0 + f\frac{r_{12}(X - X_0) + r_{33}(Y - Y_0) + r_{32}(Z - Z_0)}{+r_{13}(X - X_0) + r_{23}(Y - Y_0) + r_{33}(Z - Z_0)} + \Delta v \tag{3}$$

Here Δu and Δv are correction terms for image coordinates, r_{ij} are the components of the rotation matrix **R**. With regard to camera distortion, various models are based on either the mathematical principle, the physical principle or the mixed principle. Among many, the classical Brown model and its variants are most widely used [2]. It classifies the distortion into radial distortion Δ_r and tangential distortion Δ_t.

$$\Delta = \Delta_r + \Delta_t \tag{4}$$

In the Brown model, the radial distortion is modeled by the three parameters K_1, K_2, K_3, and P_1, P_2 are tangential distortion parameters. Furthermore, $r = \sqrt{u^2 + v^2}$:

$$\Delta \mathrm{u} = \mathrm{u}(1 + K_1 r^2 + K_2 r^4 + K_3 r^6) + \left(2P_1 \mathrm{uv} + P_2(r^2 + 2u^2)\right) \tag{5}$$

$$\Delta \mathrm{v} = \mathrm{v}(1 + K_1 r^2 + K_2 r^4 + K_3 r^6) + \left(2P_2 \mathrm{uv} + P_1(r^2 + 2v^2)\right) \tag{6}$$

Various methods based on the geometrical relationship are put forward with regard to calibration scenes, calibration models and estimation processes. In this work, experiments are mainly depending on the Matlab® Calibration Toolbox, which uses a plane chess-board.

First of all, feature points are extracted from the images, taken from different orientations of the chess-board. After the calculation of view homographies using a non-linear optimization method, extrinsic and intrinsic parameters together with distortion parameters are estimated. In the last step, all parameters are refined with a non-linear optimization as shown in (7).

$$\mathrm{E} = \sum_{i=0}^{M-1} \sum_{j=0}^{N-1} \left\| \dot{u}_{i,j} - P(\boldsymbol{a}, \boldsymbol{\omega}_i, \boldsymbol{X}_j) \right\|^2 \tag{7}$$

E is the energy function, *M* is the number of views, *N* is the number of points in each image, **a** is the vector of intrinsic parameters including distortion parameters, **ω** is the vector of view parameters, $\boldsymbol{X}_j$ are 3D coordinates of points on the calibration board, $\dot{u}$ stands for the observed image points, $P(\boldsymbol{a}, \boldsymbol{\omega}_i, \boldsymbol{X}_j)$ delivers the projected image coordinates [16].

2.1.2. Influencing Factors for Camera Calibration. In the process of calibration determination, the correlation between the calibration parameters needs to be considered. It has been found, that correlations arise between the interior parameters and exterior orientation parameters [17], [18].

In practice, good ray intersections can reduce the correlation between the exterior and interior orientation parameters. However, within the Matlab® Calibration Toolbox, the guidance on taking images is limited, as such avoiding extreme angles, changing distances, which are important while not comprehensive. Therefore, the images obtained following the instructions will not necessarily deliver satisfying calibration results. In this study, more comprehensive suggestions are given, including relative 90-degree rotations, the complete use of image format and using oblique view images.

2.1.3. Central Limit Theorem (CLT). The general idea of the Central Limit Theorem [19] is that regardless of the population distribution model, if the sample size increases, the sample mean tends to be normally distributed around the population mean, and its standard deviation shrinks as the number of samples *n* increases. The theorem holds in condition that 1) the sample size must be independent and 2) the sample size must be ''big enough''. The CLT is stated as follows:

$$\frac{\bar{X} - \mu}{\sigma/\sqrt{n}} \to N(0,1) \tag{8}$$

where $\bar{X}$ stands for sample mean, μ and σ are the mean and standard deviation of the normal distribution.

Since the extraction uncertainty of each image feature is independent, different image orientations are considered to be independent as well. Therefore, the final calibration result should fulfill the CLT, which means the parameters from succeeding calibrations should fulfill a normal distribution. More specifically, when taking a big number of images following the guidance to assume that all the images provide useful information for the calibration process, a sufficient number of random subsets can be chosen. It is assumed, that also the unstable calibration parameters should present a normal random error distribution behavior. By estimating all the calibration parameters from all calibrations, the expectation values of all parameters should be obtained in a statistical sense:

$$E(\hat{\boldsymbol{x}}) = \boldsymbol{x} \tag{9}$$

This proof is given by a simple definition of the Gauss-Markov model for estimating unknown parameters in a least-squares parameter estimation process. Let be given the linear model

$$E(\boldsymbol{l}) = \boldsymbol{Ax}; \;\; D(\boldsymbol{l}) = \sigma^2 \boldsymbol{I} \tag{10}$$

with $\boldsymbol{l}$ as observation vector of the image feature coordinates, $\boldsymbol{A}$ is the design matrix and $\boldsymbol{x}$ the vector of unknown parameters; E is the expectation and D the dispersion operator.(10) can be reformulated by using the vector of inconsistencies $\boldsymbol{v}$,

$$\boldsymbol{l} + \boldsymbol{v} = \boldsymbol{Ax}; \;\; D(\boldsymbol{l}) = \sigma^2 \boldsymbol{I} \tag{11}$$

because

$$E[\boldsymbol{l} + \boldsymbol{v}] = E[\boldsymbol{Ax}] = \boldsymbol{A}E[\boldsymbol{x}] = \boldsymbol{Ax} \tag{12}$$

with $E(\boldsymbol{l}) + E(\boldsymbol{v}) = \boldsymbol{Ax}$ if and only if $E(\boldsymbol{v}) \equiv 0$. The parameter estimation of $\|\boldsymbol{v}\|^2 = min$ finally leads to

$$\hat{\boldsymbol{x}} = (\boldsymbol{A}'\boldsymbol{A})^{-1}\boldsymbol{A}'\boldsymbol{l} \tag{13}$$

$$E(\hat{\boldsymbol{x}}) = E[(\boldsymbol{A}'\boldsymbol{A})^{-1}\boldsymbol{A}'\boldsymbol{l}] = (\boldsymbol{A}'\boldsymbol{A})^{-1}\boldsymbol{A}'E(\boldsymbol{l}) = (\boldsymbol{A}'\boldsymbol{A})^{-1}\boldsymbol{A}'\boldsymbol{Ax} = \boldsymbol{x} \tag{14}$$

2.2. Experiments

For calibration setup, a well-designed chess-board rather than a normal printed paper should be used. The feature point number should also be big enough, due to the fact, that insufficient feature points in every image will cause unstable behavior of the parameter estimation. The illumination on the image together with appropriate settings of the camera is also crucial. Those factors should be fulfilled to avoid any possible impacts. Furthermore, efforts should be paid to design a suitable image configuration.

2.2.1. Experiment Settings. Our experiments are accomplished with two mirrorless cameras, a Sony α 7R III with Zeiss Loxia 25mm lens, and a Leica Q (Typ 116) camera with Leica Summilux 28mm lens (see Figure 1). In addition, we investigate the potential of two smartphone cameras: a Galaxy Note 8 from Samsung and an iPhone 7 Plus from Apple. The Sony α 7R III and the Leica Q (Typ116) are full frame cameras, while the Galaxy Note 8 smartphone has a wide-angle camera with 1/2.55" sensor and an extra telephoto lens with 1/3.6" sensor. The iPhone 7 Plus' camera is also a wide-angle lens with 1/3″ sensor and an extra telephoto lens with 1/3.6" sensor. In this work, all experimental results are based on a special manufactured 18×29 chess-board by the company calib.io I/S (see Figure 2).

Figure. 1 Camera systems used for calibration. From left to right: Sony α 7R III, Zeiss Loxia 25mm lens and Leica Q (Typ 116) (pictures are from respective official websites).

Figure. 2 Camera calibration board (calib.io I/S).

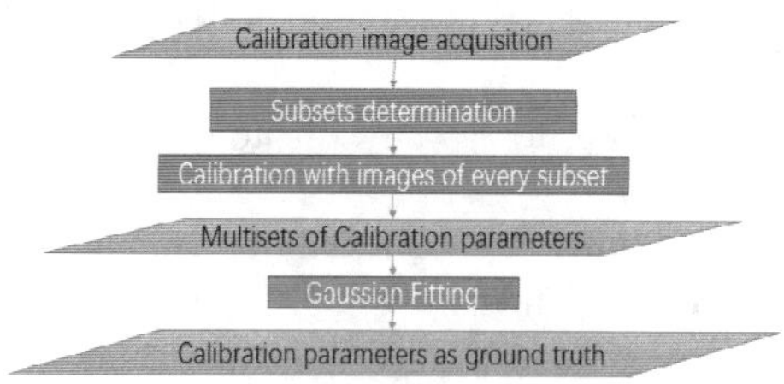

Figure. 3 Workflow.

2.2.2. Experiment Workflow. To start with, we would like to find the ground truth estimation of the intrinsic parameters. According to the instruction of OpenCV or Matlab® Calibration Toolbox [20], 10 to 20 images are appropriate for a calibration. However, in reality, images taken without enough considerations will lead to different intrinsic parameters. Therefore, in our experiment, far more than the required images are taken, with different orientations including slightly tilted or highly tilted images, rotated images, and images with different distances. If some images are ill-conditioned, they may severely influence the final calibration results. In case of ground truth estimation of intrinsic parameters, the ill-conditioned images most probably will have an impact on the confidence interval of the Gaussian fitted functions. We use the Maximum Likelihood Estimation method to calculate the expected values of the calibration parameters. The workflow for the calibration is given in Figure 3.

3. Results

3.1. Gaussian fitting for multisets of calibrations

In the experiments, different numbers of images for the subsets are tested. For a certain subset image number, 1000 subsets are chosen and thus 1000 calibrations are implemented. Therefore, every calibration parameter has 1000 results from different subsets. As stated in 2.2, a Maximum Likelihood Estimation is applied to obtain the expected values (in unit pixels). The respective pixel size of each camera is listed in Table 1, which can be used to convert the given results to metric units. Additionally, for non-full-frame sensors, the crop factor related to the sensor size is needed for the conversion to a 35mm equivalent value. The results of the focal length in X direction are displayed in Figures 4-7.

Table 1. Pixel size of the four different cameras.

Camera Type	Sony α 7R III	Leica Q	Galaxy Note 8	iPhone 7 Plus
Pixel size (μm)	4.5	5.97	1.4	1.2
Crop factor	1	1	6	7.2

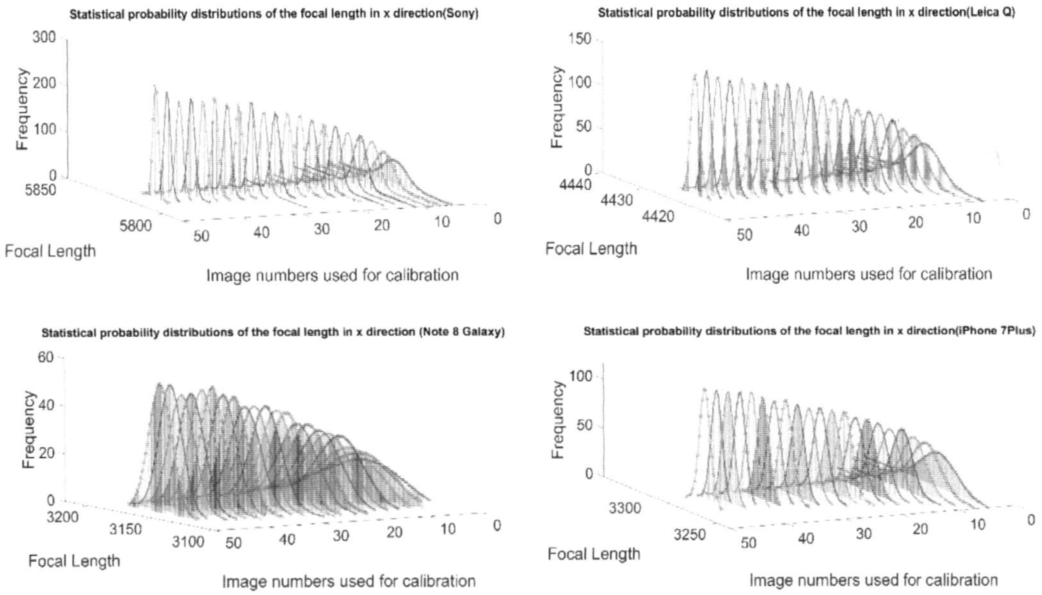

Figure. 4 Gaussian fitting experiment results. From top left to bottom right: the calibrated focal length (pixel) in x direction for the Sony α 7R III, for the Leica Q, for the Galaxy Note 8 and for the iPhone 7 Plus.

Figure 4 demonstrates that with different image subsets, the natural mean of the focal length from all the calibrations will converge. Moreover, the Gaussian fitting results indicate that more images for a subset will make a more stable calibration, though more than 20 images do not show a significant improvement. From the comparison of the four curves of Figure 5, the standard deviations of all camera calibrations converge to small values, while camera Sony α 7R III and Leica Q (Typ 116) cameras show better stability than the smartphone cameras of the Galaxy Note 8 and iPhone 7 Plus.

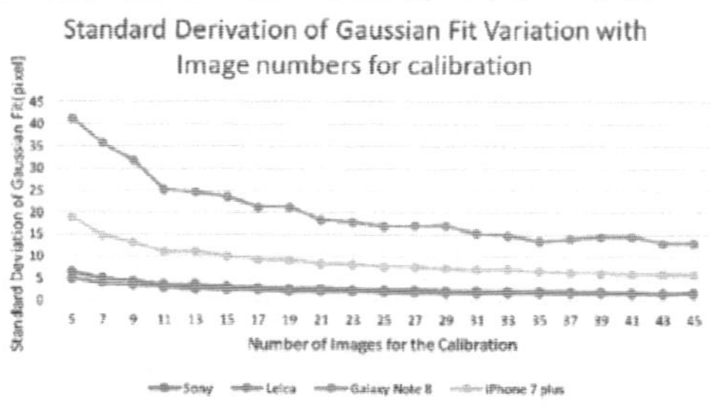

Figure. 5 Gaussian fitting standard deviations of the calibration for four types of cameras.

From the above processing, for different image numbers of subsets we obtain different Gaussian fitting curves with different expected values and standard deviations. Examples of the focal length in x direction and y direction for the four different cameras' calibration are displayed in figures 6.

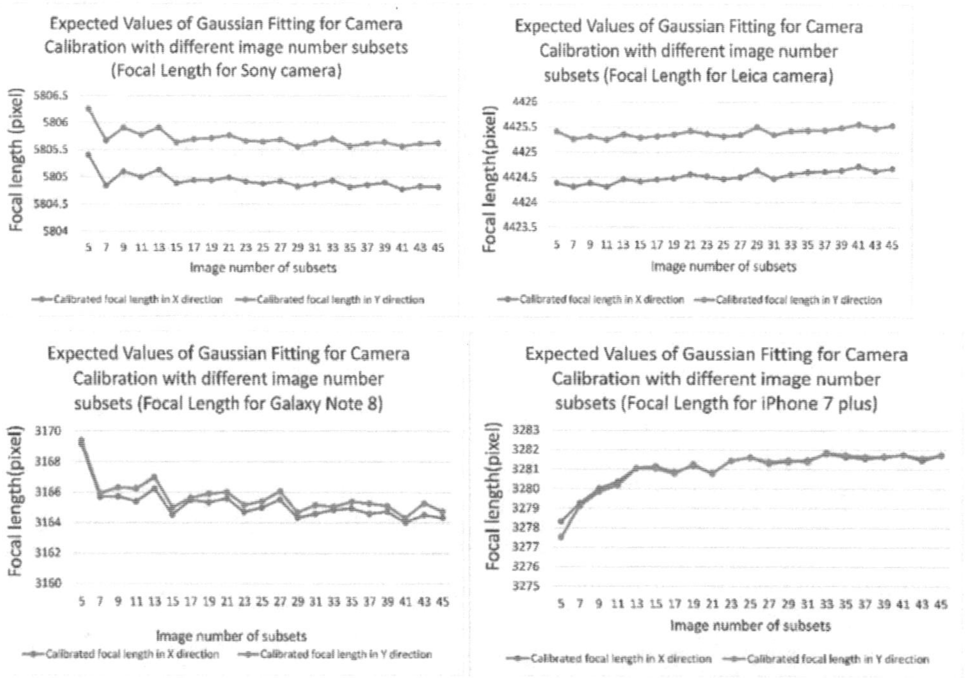

Figure. 6 Gaussian fitted standard deviations (SD) of the calibrated focal length. From top left to bottom right: SD of the Sony camera, the Leica camera, Galaxy Note 8 and iPhone 7 plus.

The outcome of the experiments above is that the calibrated parameters are stable and within small range for the calibrated focal length and the other calibration parameters. This means that the influence of the image configuration is greatly reduced by using many subsets for the calibration from a huge image bundle and then estimate the natural mean according to the Central Limit Theorem.

A weighted arithmetic mean for the expected values is determined as the final calibration parameter:

$$X = \frac{\sum_{i=1}^{n} x_i/{\sigma_i}^2}{\sum_{i=1}^{n} 1/{\sigma_i}^2} \tag{15}$$

with X as ground truth of the calibration parameter, x_i is the expected value of the ith Gaussian fitting curve, and σ_i is the corresponding standard deviation of the normal distribution. Table 2 lists the results of the individual calibration parameters.

Table 2. Calibration parameters for Sony α 7R III, Leica Q, Galaxy Note 8 and iPhone 7 Plus

Calibration parameters	Estimated Values			
	Sony	Leica	GalaxyNote8	iPhone 7 Plus
Focal length x (pixels)	5819.6096	4424.5633	3164.8576	3281.3755
Focal length y (pixels)	5819.4368	4425.4199	3165.3254	3281.3854
Principal point x (pixels)	-32.4949	3.2504	27.8190	-24.6529
Principal point y (pixels)	-1.4404	2.5041	13.6578	22.9677
Radial distortion K1	-0.0566	0.0479	0.1566	0.1951

The values of the calibration parameters from Table 2 may differ from the specifications of the vendor when transferred to mm. A reason can be the difference between the design value and the production tolerance of the product. However, the concept introduced in this paper is not suggesting readers to implement this workflow for every camera calibration, instead, the method is applied here for the estimation of a ground truth information in order to analyse the single factor influence in section 3.2 so that practical suggestions can be given for a stable calibration with about 20 images.

3.2. Image Block Analysis based on Ground Truth Estimation

After having estimated the intrinsic parameters through the Gaussian Fitting method with the whole image dataset, the composition of the images used for calibration is changed with regard to different factors for further comparison. Oblique imaging angles and roll angles (rotations along the optical axis) are important to make a strong image block for the camera calibration. Among the whole image dataset, a group of photos is taken with relatively small oblique angles, which follows the typical description of camera calibration in most open-source calibration toolboxes. Additional images with bigger oblique angles and roll angels are taken, which compose a better image block to optimize the calibration configurations. Thus, a control experiment with and without big oblique view images and roll movement images is designed to validate the suggestion for taking images. In this section, only the result of the Sony camera experiments is presented in Figure 7.

From Figure 7, the following conclusions can be made:

- big oblique view images are important for a reliable and stable camera calibration;
- weak image blocks without big oblique image and roll angle images can lead to unstable calibration results and even deviate from estimated ground truth with more images;
- roll angle images can contribute to more stable calibration results, and
- roll angle images affects more the focal length estimation than principle point;

Therefore, for non-expert users, it is of vital importance to pay more attention to the image configuration for camera calibration regarding oblique view angles as well as roll angles.

3.3. Temporal Stability Variation

We have learned that the above factors have a strong influence on the calibration. Therefore, we need to pay attention to the obtained results by taking the right images. For validation, the experiments with the Sony camera are shown as an example. Twenty independent sets of images, for which each consist of 15-25 images, are collected within one week at different epochs for camera calibration. Ten sets of images follow the typical description of camera calibration in most open-source calibration toolboxes, while another ten sets of images with the big oblique angle and roll angle images supplement the validation, so that the temporal stability of the proposed method is finally proven. Part of the resulting calibration parameters including the calibrated focal length and the principal point coordinates are shown in Figure 8.

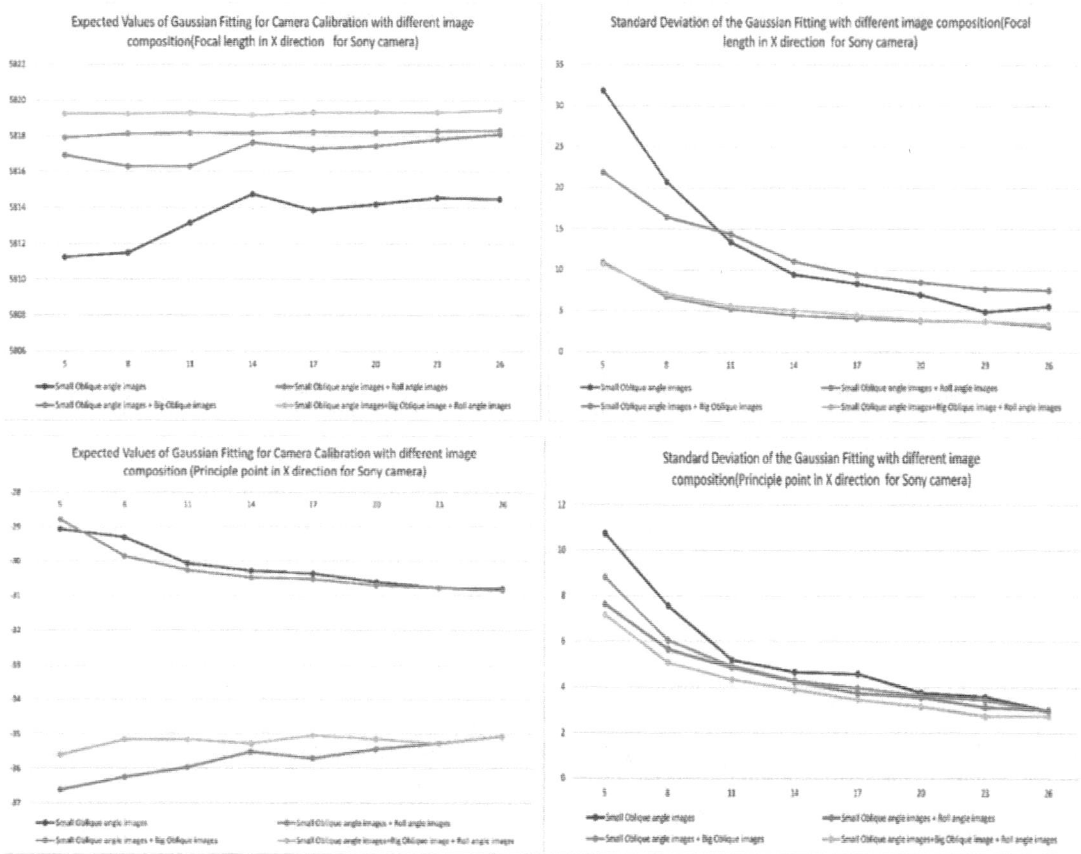

Figure. 7 Single factor analysis results of Gaussian fitting using different image compositions. From top left to bottom right: expected value for the calibrated focal length, standard deviations of the calibrated focal length, expected value for the principle point x coordinate, standard deviations of the principle point x coordinate.

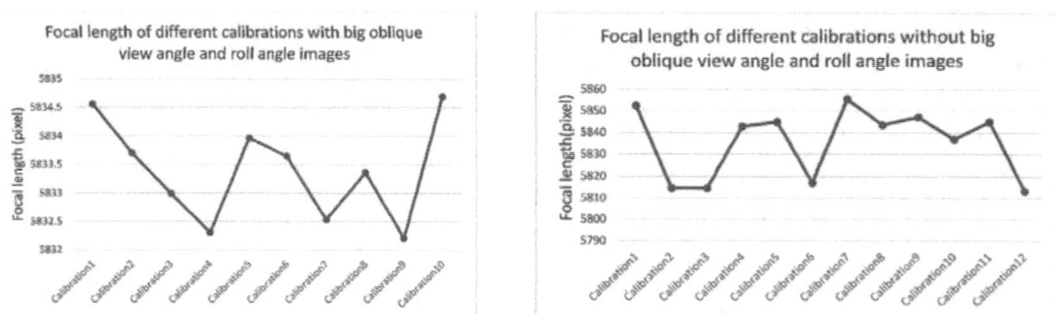

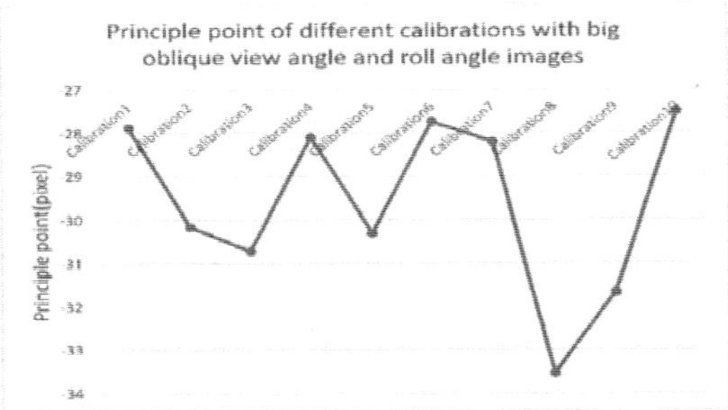

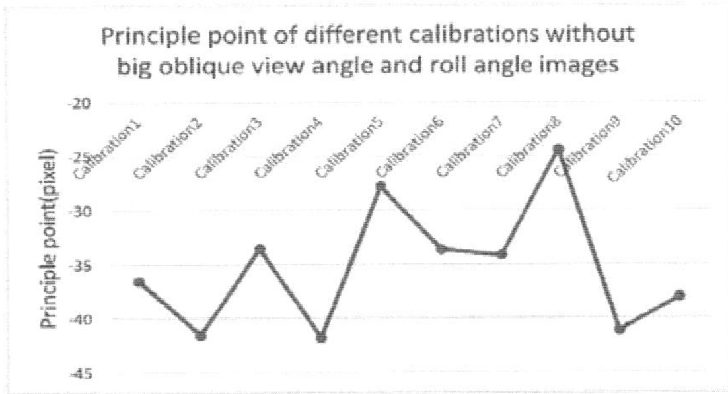

Figure. 8 Temporal Stability Variation experiment results. From top left to bottom right: focal length of different calibrations with strong image block configuration, focal length of different calibrations without strong image block configuration, principle point of different calibrations with strong image block configuration and principle point of different calibrations without strong image block configuration.

According to the figures, it can be seen that for the image datasets with big oblique angle images and roll angle images, the differences of the calibrated focal length or principal point calibration of different calibrations are within 6 pixels. However, image datasets without strong image block configurations indicate high instability with calibrated parameter deviations of up to 40 pixels. In addition, it must be noted that the experimental conditions in this section are deviating from section 3.1 regarding focus setting and other changes which might cause the deviation of the calibrated parameters between section 3.3 and previous experiments.

4. Discussions

Given the complex nature of the image acquisition process and the uncertainties of the target manufacturing, it is often difficult to identify the exact sources of errors and uncertainties and how they are combined. For reliable and repeatable calibrations for a camera with regard to operational issues especially the strategy of taking images, appropriate hardware preparations, and precisions were studied.

First of all, a sufficient number of images has to be collected, more than enough for a single camera calibration. Also the appropriate conditions concerning camera settings, hardware precision, illumination etc. should be fulfilled. Most importantly, the strategy for taking images has to be respected. Secondly, a big number of subsets from images of different configurations should randomly be chosen, and for each subset a camera calibration has to be implemented with the Matlab® Camera Calibration Toolbox. According to the Central Limit Theorem, the uncertainties of images should follow a normal distribution. Therefore, a Maximum Likelihood Estimation will deliver all calibrated camera parameters - the natural mean is regarded as ground truth of the camera. In addition, different factors, including oblique angles and 90-degree roll angles along the camera axis, should be considered by excluding the corresponding images. The results of such a process were compared and analyzed above. With knowing the influence of validated factors, during the time period of two weeks, also a normal camera calibration with 15-20 images, as suggested by most toolboxes, was repeated 10 times. Finally, the results following our guidelines show much better stability compared with calibrations only following the instructions from computer vision.

In contrast to many camera calibration simulations, our work is based on real situation experiments and aiming for a stable and repeatable camera calibration solution. Effective suggestions are given for non-experienced users. Most important for a more robust and practical solution is the study of precision figures rather than to rely just on re-projection errors, to obtain a stable and repeatable camera calibration [21].

5. Acknowledgements

The authors gratefully acknowledge the funding of this research within the Gyrolog project of the German Ministry of Education and Research (BMBF), FKZ 01UG1774X.

References

[1] Fritsch, D. „Photogrammetrische Auswertung digitaler Bilder – Neue Methoden der Kamerakalibration, dichten Bildzuordnung und Interpretation von Punktwolken“. In: Photogrammetrie und Fernerkundung, Ed. C. Heipke, Springer, Berlin, pp. 157-196, 2017.

[2] Tang, R. "Mathematical methods for camera self-calibration in photogrammetry and computer vision." Dissertation, University of Stuttgart, 2013.

[3] Tang, R., Fritsch, D. and Cramer, M. "New rigorous and flexible Fourier self-calibration models for airborne camera calibration." ISPRS Journal of Photogrammetry and Remote Sensing, Vol. 71, pp: 76-85, 2012.

[4] Tang, R., Fritsch, D., Cramer, M. and Schneider, W. “A flexible mathematical method for camera calibration in digital aerial photogrammetry”, Photogrammetric Engineering & Remote Sensing, Vol. 78, No. 10, pp. 1069–1077, 2012.

[5] Zhang, Z. “A flexible new technique for camera calibration”, IEEE Transactions on pattern analysis and machine intelligence, Vol. 22, No. 11, pp. 1330-1334, 2000.

[6] Salvi, J., Armangué, X. and Batlle, J. “A comparative review of camera calibrating methods with accuracy evaluation”, Pattern Recognition, Vol. 35, No. 7, pp. 1617–1635, 2002.

[7] Luhmann, T., Fraser, C. and Maas, H.-G. “Sensor modelling and camera calibration for close-range photogrammetry”, ISPRS Journal of Photogrammetry and Remote Sensing, Vol. 115, pp. 37–46, 2016.

[8] Abraham, S. and Hau, T. “Towards autonomous high-precision calibration of digital cameras”, Videometrics V, International Society for Optics and Photonics, Vol. 3174, pp. 82–93, 1997.

[9] Fraser, C. “Automatic camera calibration in close range photogrammetry”, Photogrammetric Engineering & Remote Sensing, Vol. 79, no. 4, pp. 381–388, 2013.

[10] Hastedt, H. and Luhmann, T. “Investigations on the quality of the interior orientation and its impact in object space for UAV photogrammetry”, International Archives of the Photogrammetry, Remote Sensing & Spatial Information Sciences, Vol. XL-1/W4, pp. 321–328, 2015.

[11] Remondino, F. and Fraser, C. “Digital camera calibration methods: Considerations and comparisons”, International Archives of Photogrammetry, Remote Sensing and Spatial Information Sciences, Vol. 36, No. 5, pp. 266–272, 2006.

[12] Fritsch, D., Wagner, J.F., Simon, S., Ceranski, B., Niklaus, M., Zhan, K., Schweizer, T. and Wang, Z. “Gyrolog – Towards VR Preservations of Gyro Instruments for Historical and Didactical Research“. In: Proceedings 2018 Pacific Neighborhood Consortium Annual Conference and Joint Meetings (PNC). Piscataway, NJ, IEEE, pp. 13-19, 2018.

[13] Taylor, J.R. “Error analysis”, Univ. Science Books, Sausalito, CA, 1997.

[14] Semeniuta, O. “Analysis of camera calibration with respect to measurement accuracy”, Procedia Cirp, Vol. 41, pp. 765–770, 2016.

[15] Peng, S. and Sturm, P. “Calibration wizard: A guidance system for camera calibration”, arXiv preprint arXiv:1811.03264, 2018.

[16] Burger, W. “Zhang’s camera calibration algorithm: In-depth tutorial and implementation”, Report, University of Applied Sciences Upper Austria, Hagenberg, 2016.

[17] Luhmann, T., Robson, S., Kyle, S. and Boehm, J. “Close-range photogrammetry and 3D imaging”. Walter de Gruyter, Berlin/Boston, 2013.

[18] Tang, R. and Fritsch, D. “Correlation analysis of camera self-calibration in close range photogrammetry”, The Photogrammetric Record, Vol. 28, no. 141, pp. 86–95, 2013.

[19] Breuer, P. and Major, P. “Central Limit Theorems for non-linear functionals of Gaussian fields”, Journal of Multivariate Analysis, Vol. 13, No. 3, pp. 425–441, 1983.

[20] Bouguet, J.-Y. ”Camera calibration toolbox for Matlab”, URL: http://www.vision.caltech.edu/bouguetj/calib_doc/, last access July 24, 2020.

[21] Fritsch, D. „Ausgleichungsrechnung damals und heute“. In: Johann Gottlieb Friedrich Bohnenberger, Ed. E. Baumann, Verlag W. Kohlhammer, Stuttgart. pp. 303–320, 2016.

Edible film coating to extend the shelf-life of fresh-cut kiwi

G. Gullifa[1]*

[1]Department of Chemistry "Sapienza" University of Rome

*E-mail: giuseppina.gullifa@uniroma1.it

Abstract. In this work, a novel coating strategy able to prolonge the shelf-life of fresh-cut kiwi is proposed and the effectiveness of the procedure was evaluated over a period of 15 days in order to propose innovative minimally processed fruits as ready-to-eat. Chitosan and carboxymethyl cellulose were compared as coating material while a number of pretreatments were tested in order to assess the most performing strategy leading to the most stable product. The complete characterization of the Volatile Organic Compounds (VOCs) was achieved by Solide Phase Micro Extraction associated to gaschromatography and mass spectrometry (SPME-GC/MS) and correlated to the weight losses occurring with the increase of aging. The obtained results suggested that the involvement of carboxymethyl cellulose and ascorbic acid on pretreated fresh-cut kiwi, permitted to extend the shelf- life of the fruit until 15 days during storage at 4°C.

1. Introduction

The demand of minimally processed fruits as ready-to-eat food is growing more and more in the last decades and innovative coating technologies are required to extend the shelf life of fresh-cut fruits [1-9]. The Kiwifruit (Actinidia deliciosa) is one of the most appreciated fruits with excellent nutraceutical properties mainly due to the high vitamin C content. In particular, fresh-cut kiwi is usually used in different salads as peeled and sliced fruits [10-16]. As a consequence, these procedures may cause the physical damage of the fruit and an increase in the ethylene production with aging. In addition, the cutting of fruits may cause the increase in the microbial content, leading to a less stable product [15, 17-24]. The application of edible film coatings is largely used to avoid contact between fruits and air and a number of procedures were tested including thermal treatments [25-32], coating [33-42] and modified atmosphere [43-46].

This study introduces a novel strategy to improve the shelf-life of fresh-cut kiwi and to ensure the appealing as well as the suitable quality of the fruits until 15 days using biodegradable cases for storage at 4°C [47-50].

2. Experimental

Materials and methods

Fresh fruits were provided by different markets in Rome and were peeled and cut into small pieces. A number of pretreatments were tested including sodium chloride (NaCl) and sodium hypochlorite (NaClO) in order to eliminate pathogenic microorganisms prior to coating.

The edible coating materials (carboxymethyl cellulose and chitosan) were purchased from Sigma-Aldrich (Milan, Italy) as powders to be dissolved in acetic acid at different ratio and the obtained solution was stirred overnight at 40°C. The coating of the fruits was performed by dipping twice the peeled fruits after pretreatment and drying with an inert gas in order to avoid oxygen inclusion during the process.

Coated and uncoated fruits were compared in order to evaluate the weight losses, expressed as percentage %ww and the chemical composition of the VOCs. To assess significances among measurements the analysis of variance (ANOVA) followed by the Tukey test was performed to compare results [50-52].

GC-MS analysis of VOCs

Head-Space Solid Phase Micro Extraction (HS-SPME) was used for the sampling of the Volatile Organic Compounds (VOCs) by a polydimethylsiloxan fiber (DVB/CAR/PDMS, 50/30 μm).

Extraction was achieved from 1 g of kiwi heated in closed vials in water bath at 50 °C for 15 minutes. The fiber was exposed for 30 minutes and then injected into the GC-MS (PerkinElmer, Waltham, MA). The chromatographic separation was obtained using an HP-5MS (30 m Å x 0.25 Å x 0.25 mm) capillary column and the carrier gas was helium at a constant flow rate of 1 mL/min. The initial temperature was set at 40°C for 1 minute, increased to 180°C at 5 °C/min and then ramped to 250°C at 10 °C/min for 5 minutes. Electron impact (EI) ionization was employed at a voltage of 70 eV and the inlet temperature was 250 °C. The temperatures of the interface, the ion source and the quadrupole were 280 °C, 230 °C and 150 °C, respectively. Mass spectra were collected in full scan mode from m/z 35 to 450 and the injection was performed in split mode with a split ratio of 60:1 [53, 54].

3. Results

In order to study the effect of two different edible coating material, fresh-cut kiwi were coated by chitosan and CMC and the weight losses as a function of the time since cutting, were calculated to evaluate the material able to maximize the respiration rate.

Chitosan and CMC were dissolved in water containing the acetic acid and stirred at 500 rpm for 24h at a constant temperature of 100 °C. Kiwifruits were peeled with an iced knife washed in NaClO and cut into pieces. A further step of washing with NaCl salt solution (0.1% w/v) and 0.1% NaClO solution or lime, was also included in order to eliminate interferences due to the microbial growth. Fruits were then dipped twice into the solution and dried under a flow of nitrogen to avoid oxygen inclusion [55-61]. Results of the weight losses are reported in Figure 1, where each coated sample with chitosan (A) or CMC (B) is compared to the corresponding reference in order to estimate the improvement due to the coating procedure.

As shown in Figure 1A, the coated kiwi fruits exhibited the lowest weight losses with respect to the reference until the ninth day, leading to a comparable total weight loss over 15 days. On the contrary, the coating of CMC ensured the highest values of weight retained from the kiwi (about 80%) and showed a significant improvement in the shelf-life of the fruit when CMC with ascorbic acid was involved after a washing step in NaCl/NaClO.

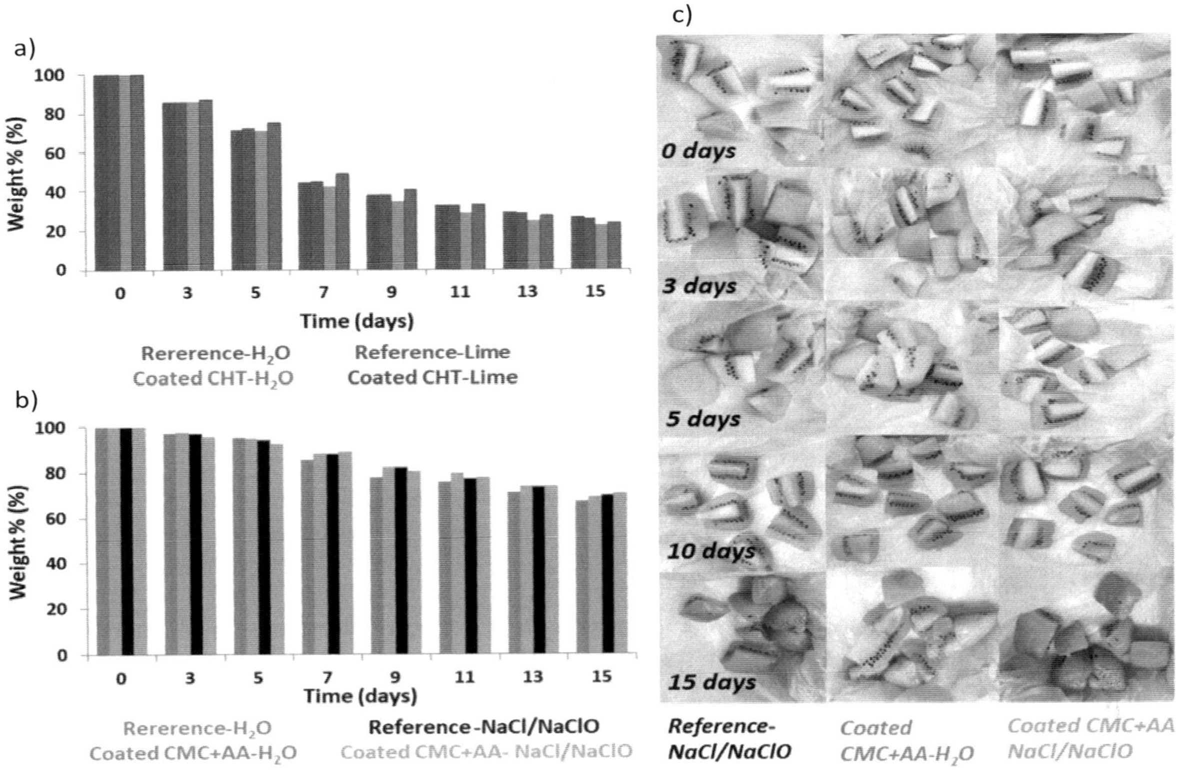

Figure. 1 a) Calculated weight losses (%) of: coated (green) and uncoated (grey) kiwi with chitosan and washed with water; coated (purple) and uncoated (red) kiwi with chitosan and washed with lime. b) Calculated weight losses (%) of: coated (blue) and uncoated (beige) kiwi with CMC and ascorbic acid and washed with water; coated (light green) and uncoated (black) kiwi with CMC and ascorbic acid and washed with NaCl/NaClO; c) visual examination of fresh-cut kiwi.

In Figure 1c, all the collected pictures of fruits over a period of 15 days are collected and compared to evaluate the effect of the coating procedure on the visual aspect of the product. The visual examination of the fruits is a critical issue for consumers that select the most agreeable and appealing product, unless its quality or declared freshness. The most performing results were obtained when fruits were washed with NaCl/NaClO and then dipped into a CMC solution with ascorbic acid.

VOCs characterization

The chemical composition of the Volatile Organic Compounds (VOCs) of fresh-cut kiwi was estimated by Headspace Micro-Extraction (HS-SPME) and the effect of the coating procedure was evaluated by comparing coated and uncoated fruits. All the recovered molecules are summarized in Figure 2a, while the behavior of each molecule was monitored over time and calculated as normalized area (Figure 2b).

a)

	Molecule	RT (min)	m/z
Alcohol	1-Penten-3-ol	2.41	57, 41
	1-Pentanol	3.98	42, 55, 70
	(E)-2-Hexen-1-ol	7.82	57, 41, 82
	1-Hexanol	8.04	56, 43, 69
	2,2-Dimethyl-1-hexanol	14.03	57, 41, 99
	2-Ethyl-1-Hexanol	16.57	57, 41, 70
	2-Nonen-1-ol	20.33	57, 41, 69
Esters	Ethyl acetate	1.81	43, 61,70
	Ethyl propanoate	2.76	57, 45, 75
	Methyl butanoate	2.92	43, 74, 87
	Ethyl isobutanoate	3.66	43, 71, 88
	Ethyl 2-methylbutanoate	6.89	57, 102, 41
	Ethyl benzoate	23.54	105, 77, 122
Aldehydes	Pentanal	2.57	44, 58
	2-methyl-4-pentenal	4.8	41, 56, 69
	(E)-2-Hexenal	7.01	41, 55, 69
	Heptanal	9.9	70, 41, 55
	Octanal	15.07	43, 56, 84
	Nonanal	17.68	57, 41, 98
	Decanal	20.03	43, 57,70
Terpenoids	α-Pinene	10.93	93, 77, 41
	β-Pinene	13.26	93, 41, 69
	Limonene	16.16	68, 93, 79

b)

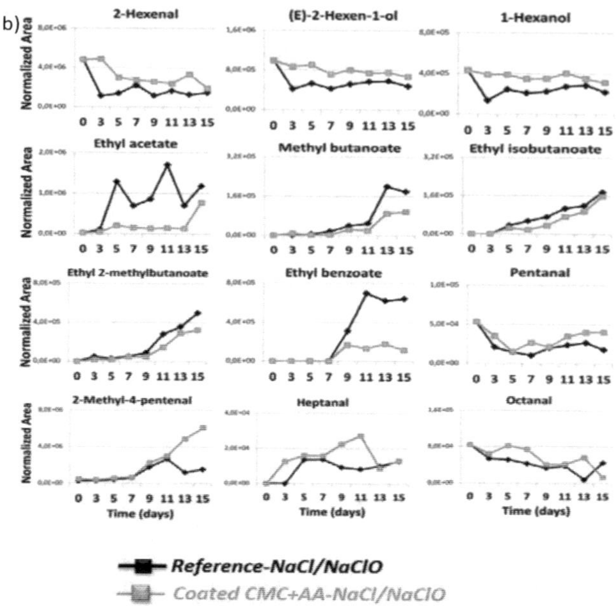

Figure. 2 a) Volatile Organic Compounds (VOCs) recovered in fresh-cut kiwi and the monitoring of each molecule over time for the reference uncoated kiwi (black) and the coated with CMC and ascorbic acid after washing in NaCl/NaClO.

As reported in Figure 2b, the aroma flavour of kiwi is strongly affected by the ripening and the aging of fruits after cutting. Significantly higher values of esters (ripening index) may be recovered in the reference samples compared to the coated fruits, suggesting positive effect of the coating material in preserving the aroma and thus the quality of the kiwi.

On the contrary, aldehydic compounds were found to be higher in the coated samples with respect to the reference ones. This behavior confirms that the flavour was most preserved in the coated fruits due to the coating process.

Conclusions

This study introduces a novel way to perform coating processes by edible film materials to extend the shelf-life of fresh-cut kiwi. The assessment of the coating conditions was performed by calculating the weight losses of fruits, the relative abundance of the molecules recovered in the fresh-cut kiwi and by examining the visual aspect of the fruits over a period of 15 days at 4°C. The comparison of the coated and uncoated fruits demonstrated that the involvement of CMC and ascorbic acid on pretreated fresh-cut kiwi, can improve the quality of the fruit.

References

[1] Dia T, Biliaderis CG, Gerasopoulos D, Sfakiotakis E, Fisk CL, Silver AM, Strik BC, Zhaoa Y 2008 *Postharvest Biology and Technology* **47**, 338–345

[2] Napoli A, Aiello D, Di Donna L, Sajjad A, Perri E, Sindona G 2006 *Anal. Chem.* **78 (10)**, 3434-3443.

[3] Reddy C E, Albanito L, De Marco P, Aiello D, Maggiolini M, Napoli A, Musti A M 2013 *Cell Death and Dis.* **4 (10)**, e852.

[4] Napoli A, Aiello D, Aiello G, Cappello M S, Di Donna L, Mazzotti F, Materazzi S, Fiorillo M, Sindona G 2014 *J. Prot. Res.* **13 (6)**, 2856-2866.

[5] Cardiano P, Giuffre O, Napoli A, Sammartano S 2009 *New J Chem* **33 (11)**, 2286-2295.

[6] Materazzi S, Finamore J, Risoluti R, Napoli A 2014 *Microchem. J.* **115**, 27-31.

[7] Jahouh F, Saksena R, Aiello D, Napoli A, Sindona G, Kováč P, Banoub J H *J. Mass Spectrom.* **45 (10)**, 1148-1159.

[8] Napoli A, Athanassopoulos C M, Moschidis P, Aiello D, Di Donna L, Mazzotti F, Sindona G 2010 *Anal. Chem.* **82** **(13)**, 5552-5560.
[9] Aiello D, Cardiano P, Cigala R M, Gans P, Giacobello F, Giuffrè O, Napoli A, Sammartano S 2017 *J Chem Eng Data* **62** **(11)**, 3981-3990.
[10] Xu S, Chen X, Sun DW 2001 *Journal of Food Engineering* **50**, 211-216.
[11] Aiello D, Casadonte F, Terracciano R, Damiano R, Savino R, Sindona G, Napoli A 2016 *Oncoscience* **3** **(7-8)**, 220-241.
[12] Materazzi S, Peluso G, Ripani L, Risoluti R 2017 *Microchem. J.* **134**, 277-283.
[13] Sergi M, Gentili A, Perret D, Marchese S, Materazzi S, Curini R 2007 *Chromatographia* **65**, 757-761.
[14] Aiello D, Furia E, Siciliano C, Bongiorno D, Napoli A 2018 *J. Mol. Liq.* **269**, 387-397.
[15] Bartella L, Mazzotti F, Napoli A, Sindona G, Di Donna L 2018 *Anal. Bioanal. Chem.* **410** **(8),** 2193-2201.
[16] Aiello D, Siciliano C, Mazzotti F, Di Donna L, Athanassopoulos C M, Napoli A 2018 *RSC Adv.* **8** **(63)**, 36104-36113.
[17] Beirão-da-Costa S, Steiner A, Correia L, Leitão E, Empis J, Moldão-Martins M 2008 Eur Food Res Technol **226**, 641–651.
[18] Aiello D, Siciliano C, Mazzotti F, Di Donna L, Athanassopoulos C M, Napoli A 2020 *Food Chem.* **307**, 125527.
[19] Imbrogno S, Aiello D, Filice M, Leo S, Mazza R, Cerra M C, Napoli A 2019 *Sci. Rep.* **9** **(1)**, 18953.
[20] Chillè D, Aiello D, Grasso G I, Giuffrè O, Napoli A, Sgarlata C, Foti C 2020 *J Environ Sci (China)* **94**, 100-110.
[21] Salvatore L, Gallo N, Aiello D, Lunetti P, Barca A, Blasi L, Madaghiele M, Bettini S, Giancane G, Hasan M, Borovkov V, Natali M L, Campa L, Valli L, Capobianco L, Napoli A, Sannino A 2020 *Int. J. Biol. Macromol.* **154**, 291-306.
[22] Furia E, Napoli A, Tagarelli A, Sindona G 2013 *J. Chem. Eng. Data* **58** **(5)**, 1349-1353.
[23] Bartella L, Di Donna L, Napoli A, Siciliano C, Sindona G, Mazzotti F 2019 *Food Chem.* **278**, 261-266.
[24] Di Donna L, Benabdelkamel H, Mazzotti F, Napoli A, Nardi M, Sindona G 2011 *Anal. Chem.* **83** **(6)**, 1990-1995.
[25] Benítez s, Achaerandio I, Sepulcre F, Pujolà M 2013 *Postharvest Biology and Technology* **81**, 29–36.
[26] Napoli A, Aiello D, Di Donna L, Moschidis P, Sindona G 2008 *J. Prot. Res.* **7** **(7)**, 2723-2732.
[27] Bartella L, Di Donna L, Napoli A, Siciliano C, Sindona G, Mazzotti F 2019 *Food Chem.* **278**, 261-266.
[28] De Angelis Curtis S, Kurdziel K, Materazzi S, Vecchio S 2008 *J. Therm. Anal. Calorim.* **92**,109-114.
[29] Risoluti R, Gullifa G, Fabiano MA, Sorrentino F, Caprari P, Materazzi S 2018 *J. Therm. Analys. Calorim.* **134**,1299–1306.
[30] Risoluti R, Gullifa G, Battistini A, Materazzi S 2019 *Anal. Chem.* **91(10)**, 6435-6439.
[31] Migliorati V, Ballirano P, Gontrani L, Materazzi S, Ceccacci F, Caminiti R 2013 *J. Phys. Chem. B* **117**, 7806-7818.
[32] Aiello D, Materazzi S, Risoluti R, Thangavel H, Di Donna L, Mazzotti F, Casadonte F, Siciliano C, Sindona G, Napoli A 2015 *Mol. BioSyst.* **11**, 2373-2382.
[33] Hua C, Li Y, Wang X, Kai K, Su M, Shi W, Zhang D, Liu Y 2019 *Scientia Horticulturae* **246**, 700–709.
[34] Agas D, Marchetti L, Menghi G, Materazzi S, Materazzi G, Capacchietti M, Hurley M M, Sabbieti M G 2008 *J. Cell. Physiol.* **214**, 145-152.
[35] Gentili A, Caretti F, D'Ascenzo G, Mainero Rocca L, Marchese S, Materazzi S, Perret D 2007 *Chromatographia* **66**, 669-676.
[36] Vecchio S, Materazzi S, Wo L W, De Angelis Curtis S 2013 *Thermochim. Acta* **568**, 31-37.

[37] Materazzi S, De Angelis Curtis S, Vecchio Ciprioti S, Risoluti R, Finamore J 2014 *J. Therm. Anal. Calorim.* **116**, 93–98.
[38] Materazzi S, Risoluti R, Napoli A 2015 *Thermochim. Acta* **606**, 90–94.
[39] Risoluti R, Gullifa G, Battistini A, Materazzi S 2019 *Drug and Alcohol Depend* **205**, 107578.
[40] Foglia P, Vuono D, Siciliano C, Napoli A, Nagy J B, De Luca P 2019 *Iop Conf. Ser. Mater. Sci. Eng.* **572(1)**, 012047.
[41] Materazzi S, Vecchio S, Wo L W, De Angelis Curtis S 2012 *Thermochim. Acta* **543**, 183-187.
[42] Aiello D, Siciliano C, Mazzotti F, Donna L D, Risoluti R, Napoli A 2020 *Molecules* **25 (7)**, 1485.
[43] Rocculi P, Romani S, Dalla Rosa M 2005 *Postharvest Biology and Technology* **35**, 319–328.
[44] Risoluti R, Gregori A, Schiavone S, Materazzi S 2018 *Anal. Chem.* **90**, 4288-4292.
[45] Materazzi S, Risoluti R, Pinci S, Romolo F S 2017a *Talanta* **174**, 673-678.
[46] Temperini A, Aiello D, Mazzotti F, Athanassopoulos C M, De Luca P, Siciliano C 2020 *Molecules* **25 (6)**, 25061313.
[47] Risoluti R, Fabiano M A, Gullifa G, Vecchio Ciprioti S, Materazzi S 2017 *Appl. Spectrosc. Rev.* 52, 39-72.
[48] Materazzi S, Foti C, Crea F, Risoluti R, Finamore J 2014 *Thermochim. Acta* **580**, 7–12.
[49] Navarra M A, Materazzi S, Panero S, Scrosati B 2003 *J Electrochem Soc* **150 (11)**, A1528-A1532.
[50] Aiello D, Giambona A, Leto F, Passarello C, Damiani G, Maggio A, Siciliano C, Napoli A 2018 *Sci. Rep.* **8 (1)**, 10973.
[51] Materazzi S, Napoli A, Risoluti R, Finamore J, D' Arienzo S 2014 *Int. J. Mass Spectrom.* **365-366**, 372-376.
[52] Marchetti L, Sabbieti M G, Menghi M, Materazzi S, Hurley M M, Menghi G 2002 *Histol. Histopathol.* **17 (4)**, 1061-1066.
[53] Kurdziel K, Glowiak T, Materazzi S, Jezierska J 2003 *Polyhedron* **22 (23)**, 3123-3128.
[54] Papadopoulos C, Cristvo B, Ferenc W, Hatzidimitriou A, Vecchio Ciprioti S, Risoluti R, Lalia-Kantouri M 2016 *J. Therm. Anal. Calorim.* **123**, 717–729.
[55] Romolo F S, Ferri E, Mirasoli M, D' Elia M, Ripani L, Peluso G, Risoluti R, Maiolini E, Girotti S 2015 *Forensic Sci Int.* **246**, 25-30.
[56] Catauro M, Naviglio D, Risoluti R, Vecchio Ciprioti S 2018 *J. Therm. Anal. Calorim.* **133**, 1085–1092.
[57] Risoluti R, Materazzi S, Sorrentino F, Bozzi C, Caprari P 2018a *Talanta* **183**, 216–222.
[58]Materazzi S, Risoluti R 2014 *Applied Spectroscopy Reviews*, **49 (8)**, 635-665.
[59]Crea F, Falcone G, Foti C, Giuffrè O, Materazzi S 2014 *New Journal of Chemistry*, **38**, 3973-3983
[60]Risoluti R, Piazzese D, Napoli A, Materazzi S. 2016 *Journal of Analytical and Applied Pyrolysis*, **117**, 82-87
[61]Bretti C, Crea F, De Stefano C, Foti C, Materazzi S, Vianelli G. 2013 *Journal of Chemical and Engineering Data*, **58**, 2835-2847

Seismic performance of reinforced concrete tall buildings with conventional and non-conventional construction systems

L Arita[1], R Lezameta[1], G Huaco[2], D Garber[3]

[1] Bachelor, Peruvian University of Applied Science UPC, Lima Peru
[2] Professor, Peruvian University of Applied Science UPC, Lima Peru
[3] Associate Professor, Florida International University FIU, Miami Florida USA

E-mail: pccighua@upc.edu.pe

Abstract. Currently in the city of Lima there is a limited number of high-rise buildings. Therefore, there is not much literature on this type of building in Peru. Peruvian codes focus on medium and low-rise buildings. For this reason, studies are required to analyze and design these tall buildings more appropriately according to the reality of the country. In this article, a pushover modal analysis of 6 types of 35-Story reinforced concrete buildings in the city of Lima will be developed. Three building models with different structural systems and square and rectangular plan are proposed, being the areas of 29m x 29m and 52m x 26m respectively. These structural systems are rigid core and frames with an energy dissipation system (fluid viscous dampers and shear-link-bozzo dissipators SLB) in order to study their behavior against seismic stresses. These buildings were based on the criteria and requirements of the current codes in the country as well as the distribution of the floor plan of buildings commonly used for offices and homes. Natural periods (T) were found to range from 2.6 to 3.3 seconds for rigid core buildings. There is an increase for viscous damping buildings from 4.2 to 5.4 seconds and also for SLB devices to range from 3.7 to 4.6 seconds. In turn, modal static nonlinear analysis was performed to obtain the capacity curves for each type of building, which were compared with the seismic demands according to the design provisions of the Peruvian seismic standard E.030 and an average of design spectra. of acceleration records of severe seismic events in Peru and scaled in a range of 0.2T to 1.5T. The performance points for each building case were determined following the ATC-40 methodologies, finding that tall buildings with a rigid core have approximately twice the stiffness of buildings with SLB dampers, as well as low ductility, unlike buildings with dissipators, that have a high ductility.

1. Introduction.

There are currently only a few buildings in Lima taller than 30 stories; some examples are the Westin Lima Hotel & Convention Center (30 stories), the "Banco de la Nación" Tower (30 stories), and, the most current, MET project (37 stories). Current Peruvian design codes are calibrated primarily for medium and low-rise buildings. Because construction of high-rise buildings is already underway, studies should be carried out to determine required updates to the Peruvian design codes to account for the different behavior of these types of buildings [1] [2] [3].

For the correct analysis of the behavior of high-rise buildings, different types of non-linear analysis have been performed, among which are the most well-known non-linear static analysis (Pushover) and incremental dynamic analysis (IDA). That second one is the most accurate analysis, but it requires a large computational load at the time of performing the analysis. Therefore, different variations of static non-linear analyses have been developed that give similar results to dynamic analyses but in a shorter computational time and analysis load [4] [5] [6].

In this study, a pushover modal analysis has been used, since it gives very similar results to the actual behavior of tall buildings. This analysis has been applied to a total of 6 regular buildings of 35 stories being total of 133m height, of which 3 are square plan and 3 are rectangular plan. In addition, each of the buildings on each floor has a different type of construction system: rigid core, viscous fluid dampers and SLB (Shear Link Bozzo) dissipators. Once these analyses were carried out, the capacity curves of each building were obtained. These curves were used to obtain the performance points and quantify their vulnerability.

2. Structure Capacity vs. Seismic Demand

Based on the collected plans, the dimensions of 29x29 m were defined for the square plan and 52x26 m for the rectangular plan, as shown in Figure 1. The dimensions for the structural elements in the first story were 50 cm thick for the structural walls, 80 x 80 cm columns for rigid core buildings, viscous fluid damper and SLB dissipators for the square plan. For the rectangular plan, the columns were 80 x 80 cm for a rigid core and 100 x 100 cm for a viscous fluid damper and SLB dissipators. The beams sections were 45x90 cm and 50x90 cm, they for square and rectangular plan respectively for every structural system at firsts levels. These dimensions decrease throughout the upper levels. Values of f'c vary from 70MPa for first levels to 42Mpa for upper levels.

(a) 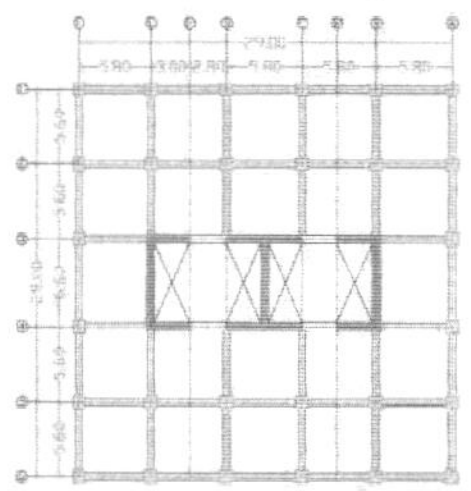 (b) 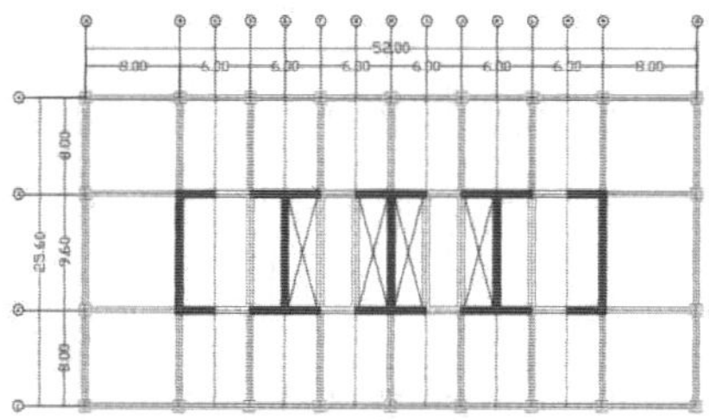

Figure. 1 (a) Square plan detail. (b) Rectangular plan detail

2.1. Seismic analysis and structural design

For Rigid Core Buildings, according to the Peruvian Code E.030 [7], seismic parameters were Z = 0.45, U = 1 common use such as apartments or offices, S = 1 considering that this type of infrastructure has deep foundation based on rock, and reduction value R = 6 for structural walls system. Minimum C / R = 0.11 was used because of large natural periods of the tall buildings. Then, the Seismic Coefficients obtained was 5%, and minimum shear base forces were used to meet E030. The structural design was carried out according to the indications of the Peruvian Code E.060.

For Buildings with viscous fluid dampers, the structural system was changed to a frame system provided with the aforementioned dimensions. Besides it was used 4 pairs of ground motion records from severe earthquakes in Peru: Lima 1966, 1970, 1974 and Pisco 2007. These were scaled based on a spectrum of pseudo accelerations with R = 1 in the SeismoMatch 2016 program in a range of 0.2T to 1.5T according to Peruvian Code E030. Then, a time-history analysis was carried out with each of the earthquakes previously scaled and thus obtain a range of floors where the maximum drifts are observed so that viscous fluid dampers can be placed in them [8], being for both floors 16 to 32. For the calculation and design of these non-linear viscous fluid dampers, an objective drift was determined according to the HAZUS and SEAOC tables [9], and an effective damping was calculated for each axis of each plant that are within a range of 20 to 40% by manufacturer's recommendation [10]. In order to define this type of shock absorbers, the damping C value varies from 200t-s/m to 1200t-s/m, the stiffness of the metal arm the K value from 24470t/m to 33010t/m and coefficient $\alpha = 0.5$. Figure 2 shows the distribution of dampers in buildings.

For SLB dissipators buildings, the same structural frame used for fluid viscous case was used. But, in this case, the structural design of the structural elements of buildings was first carried out using design combinations that only present gravity forces [11]. The SLB devices were located from floors 1 to 33 on both plans as can be seen in Figure 3. Once the dissipators were placed in the buildings, the pseudo-acceleration spectrum for the E.030 code was included to be able to design the devices. The DISSIPA-SLB 2019.2 [12] plugin was used, which serves to define the dissipators as "links" elements, then collect the shear that reaches the device due to the load combinations, including the effects of the earthquake, and finally, to propose a type of SLB dissipators that supports the loads submitted having a demand-capacity factor (D / C) not greater than 1.5. Once the types of devices complying with the aforementioned procedure were obtained, the drift verification was performed and then a nonlinear time-history analysis was carried out to verify the hysterical behavior of the dissipators.

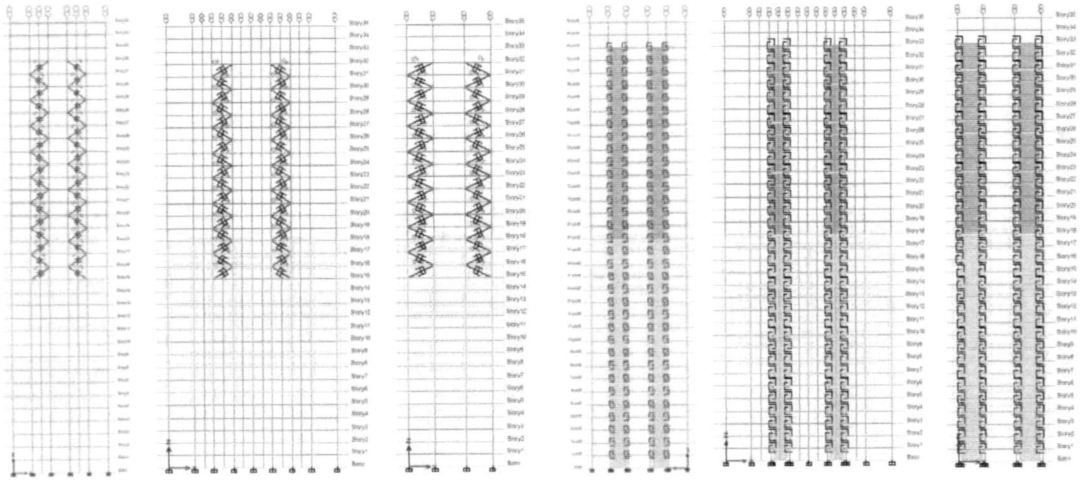

(a) (b) (a) (b)

Figure. 2 Location of viscous fluid dampers - left, and SLB dissipators – right: Square Plan longitudinal and transversal direction (a) Rectangular Plan longitudinal and transversal direction (b)

Maximum drift reached per orthogonal direction of the building were from 0.28% to 0.46% for Rigid Core, 0.42% to 0.53% for Fluid Viscous and 0.26 to 0.45% for SLB structures, being less than the limit required by the Peruvian Code 0.70%. Besides, for core walls it was designed horizontal reinforcement ratio between 0.28% to 0.40% for shear forces, and vertical reinforcement ratio of 0.28% for axial-bending forces, all of them larger than minimum 0.25% ratio as E060 Peru Concrete Design Code [13]. For columns reinforcement, it was found ratios of 1.28% for rigid core buildings, 3.58% for Fluid Viscous and SLB disipators, every case larger ratio of 1% as E060 Code requests.

3. Modal Pushover Analysis – Capacity Curves and determination of Seismic Demand Curves

The buildings were modeled again using a pushover modal analysis to determine the capacity curves of each. Beams and columns were modeled as frame elements in all the buildings; structural walls in buildings with a rigid core were modeled as frame elements having an infinitely rigid beam at the top of these [14]. The nonlinear load cases were defined to have a combination of 1.1 CM + 0.25 CV and standard point loads applied at the center of mass of each floor of the buildings. These load patterns have a modal distribution according to the mode that corresponds to each axis that was evaluated. The reinforcement obtained per each building was used to calculate the moment-curvature diagrams according to predetermined tables for each type of element from ASCE41-16[15]. Plastic hinges (M3) were assigned at a relative distance of 5% and 95% on the beams of all the buildings. Plastic hinges with biaxial behavior (P-M2-M3) at the same relative distance were assigned for the columns, and plastic hinges of uniaxial behavior were assigned for the plates according to the major axis (P-M2 and P-M3) at a relative distance of 50%. Subsequently, the control point was defined in the center of mass of the last story of each building and with a control distance that was iterated until the collapse of each one. Finally, pushover modal analyses were performed for each axis of each building obtaining the capacity curves (basal shear vs. roof displacement). It can be seen that buildings have different types of behavior, evidencing that buildings with a rigid core have greater rigidity in the linear range and that buildings with an energy dissipation system are more ductile and allow greater displacement before failure. Figure 3 shows the Capacity Curve per building case.

Two types of demand curves were considered. First one by the design spectrum according E.030 Peruvian Code, and second one by the average of Sa vs. T spectrum from of 4 pairs scaled ground motion records by SeismoMatch 2016 [16]. The ADRS conversions according to ATC-40 provisions[17] were applied to obtain Sa vs Sd and Shear Base vs Displacement curves

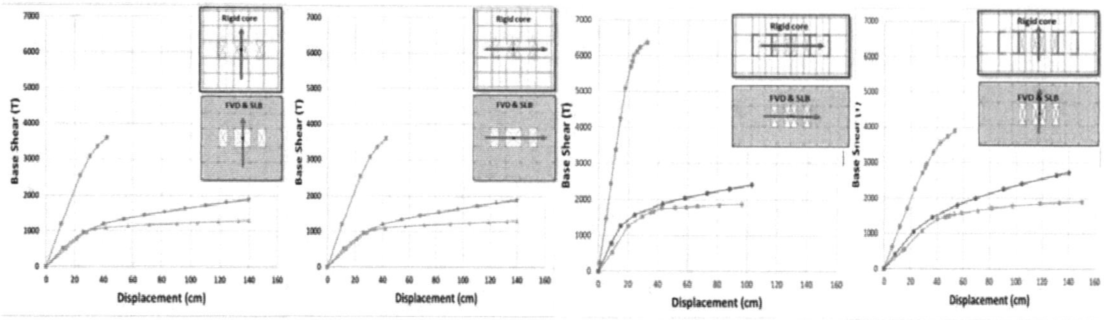

Figure. 3 Pushover Curves per case

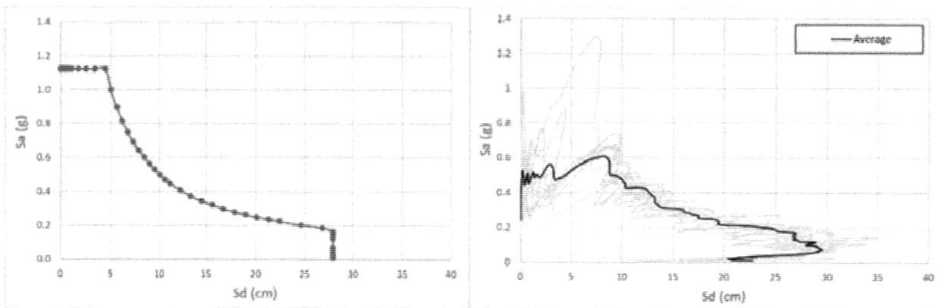

Figure. 4 Demand Spectrum Sd vs Sa of the E.030 code and average of 4 pairs of ground motion records on the Y axis of the square plan

3.1. Determination of the seismic performance point

The performance point is obtained by superimposing and intersecting the capacity curves of the buildings with their respective demand spectrum. Where, the Peruvian Code gives us a spectrum of elastic demand that must be reduced by a factor R = 6 for rigid cores, R=8 for fluid dampers and R=6 for SLB. The displacement would be obtained from the intersection of the capacity curve with the elastic design demand spectra of the E.030 standard (R = 1) and the base shear force (V) by intersecting the capacity curve with a spectrum of inelastic demand divided by a reduction factor R [13]. Results for some comparisons between capacity and demand curves are shown in figures 5 and 6.

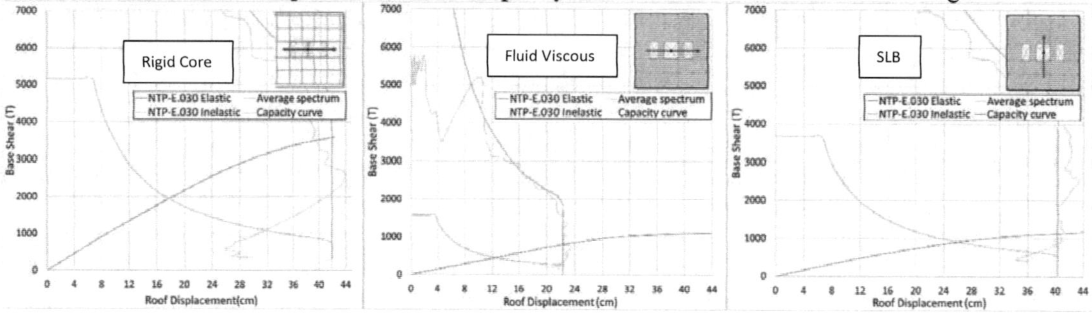

Figure. 5 Performance Points - Square Floor Plan

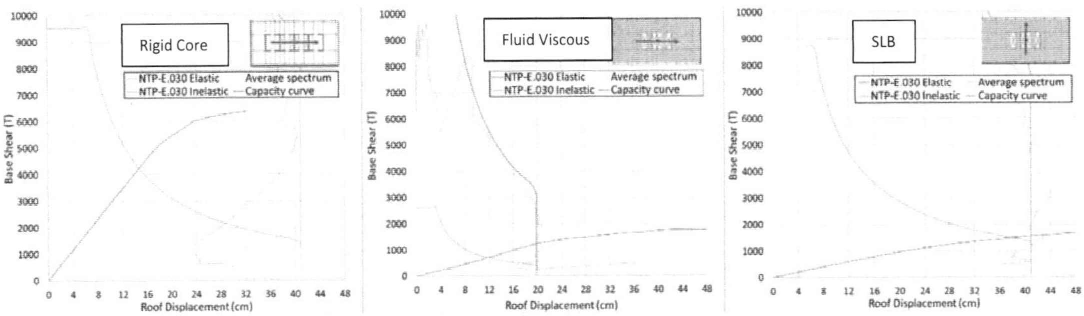

Figure. 6 Performance points – Rectangular Floor Plan

4. Conclusions

In the obtained pushover curves, it can be seen that the buildings with a rigid core have a greater stiffness in the linear range and less ductility in the non-linear range compared to the buildings with dissipation systems.

The collapse mechanism for buildings with a rigid core was caused by a flexural failure of the structural walls in the first floors; therefore, only plastic hinges are formed with flexure rather than shear. The failure that can be seen is fragile in nature. The collapse mechanism in buildings provided with viscous fluid dampers and SLB dissipators is ductile failure with the formation of plastic hinges at the base of the columns.

The seismic performance of rigid core buildings according to the explained methodology was obtained in the linear range of the structure's capacity curve. The performance drifts are within the limit required by the Peruvian Code. However, these buildings would not comply with performance requirements, since they have small drifts in the non-linear range. These small non-linear drives would generate collapse based on a demand curves using the seismic Peruvian code. For the buildings with viscous fluid dampers, the seismic performance was obtained in the limit of the non-linear range. This would lead to the structure forming plastic hinges in the beams based on a seismic design using the Peruvian Code. Finally, in the case of buildings with SLB dissipators, the seismic performance was obtained in the non-linear range, causing the structure to present plastic hinges in the beams in most of the floors, but without collapse.

5. References

[1] Gonzales, S., Herbozo, L., & Huaco, G. (2019) Approach of Posttensioned Concrete Slab Thickness Comprising the Structural Lateral System of High-rise Buildings with Rigid Core. XVII LACCEI International Multi-Conference For Engineering, Education And Technology - The OAS Summit of Engineering for the Americas, July 2019 DOI: http://dx.doi.org/10.18687/LACCEI2019.1.1.107

[2] Quezada, E., Serrano Y., & Huaco, G. (2019) Dynamic Amplification Factor Proposal for Seismic Resistant Design of Tall Buildings with rigid core structural system. Proceedings of the 5th Brazilian Technology Symposium - Emerging Trends, Issues, and Challenges in the Brazilian Technology, Volume 2, Smart Innovation, Systems and Technologies, Springer.

[3] E. N. Quezada Ramos, Y. Serrano Arone and G. Huaco, "A comparative study of the seismic base shear force and story drift ratios using Time History and Modal Spectrum Analysis according to Peru Code E.030 and ASCE 7.16 on high-rise buildings," 2020 Congreso Internacional de Innovación y Tendencias en Ingeniería (CONIITI), Bogotá, Colombia, 2020, pp. 1-5, DOI: 10.1109/CONIITI51147.2020.9240330.

[4] Saisaran, G. S., Prasad, V. Y. D., & Das, T. V (2016), Push Over Analysis for Concrete Structures at Sesimic Zone-3 using Etabs Software, International Journal of Engineering Research, 5 (03).

[5] Chopra, A. K (1995), Dynamics of structures: theory and applications to earthquake engineering, Prentice Hall, USA

[6] Liu, Y., & Kuang, J. S (2019), Estimating seismic demands of singly symmetric buildings by spectrum-based pushover analysis. Bulletin of Earthquake Engineering, 17 (4), 2093-2113.
[7] National Building Regulations, Peruvian Technical Building Standard E.030 "Earthquake-resistant Design" 2018, Peru.
[8] Castro, G. V., & Sarmiento, R. O (2009), Buildings with energy dissipators, National Assembly of Rectors, Lima Peru
[9] HAZUS (2003), Federal Emergency Management Agency FEMA, Mutihazard Loss Estimation Methodology Hazus, Washington, D.C.
[10] Taylor, D. P (1999), Buildings: design for damping, Proceedings of the Boston Society of Civil Engineers, BSCES, Lecture Series, "Dynamics of Structures", USA.
[11] Gonzales, H., Pantoja, M., & Herrera, G. Analysis And Design Using Slb Type Seismic Dissipators Application Examples, Version 04-06-2019, Luis M. Bozzo Collaborators.
[12] Bozzo, L et. to the. (2019), Procedure manual for design with SLB.
[13] National Building Regulations, Peruvian Technical Building Standard E-060 "Reinforced Concrete" 2009
[14] Garcia Murcia, D. A (2015), Pushover Analysis in Court Walls: State of The Art and Application (Doctoral dissertation, Industrial University of Santander, School of Civil Engineering).
[15] ASCE / SEI 41-16 (2017), Seismic Evaluation and Retrofit of Existing Buildings, Reston, Virginia: American Society of Civil Engineers.
[16] SEISMOMATCH v.2018, SeismoSoft Earthquake Engineering Software, Italy
[17] ATC-40 (1996), Seismic Evaluation and Retrofit of Concrete Buildings, ATC-40 Report, Volumes 1 and 2, California: Applied Technology Council

An assessment of some physical properties of different brands of cement in Nigeria

Munachiso C. Ogbodo[1] and Akpabot Ifiok Akpabot[2*]

[1]Department of Civil Engineering, Nigeria Maritime University, Okerenkoko, Nigeria
[2]Department of Civil Engineering, Covenant University, Ota, Nigeria

E-mail: munachiso.ogbodo@nmu.edu.ng

Abstract. The durability of infrastructures has become a significant issue starring the face of professionals in the building industry as most recently constructed structures are deficient in this regard. This study was carried out to assess the physical properties of five different brands of Portland cement commonly used in Nigeria. The specific surface method was adopted in obtaining the fineness, and the specific gravity bottle was used to determine the specific gravity of the different cement brands. The result shows that sample A had the highest fineness value of 360m2/kg and the only sample that met the ASTM standard value of 280 m2/kg. This implies that Sample A has a greater surface area for hydration which should lead to faster development of early strength and a higher rate of heat evolution. Sample E had the least fineness value of 168 m2/kg. This means that Sample E has the least surface area for hydration and should be slow in early strength gain. Sample B had the highest specific gravity value of 3.266 and the highest value of bulk density of 1462.085kg/m3. The implication is that Sample B has more moisture because it has the highest specific gravity value. The quantity of sample B that will be required in the concrete making will be less relative to other samples as it had the highest bulk density.

1. Introduction

Most developed nations rely on the construction industry as a significant contributor to the growth of their economies. In developing societies, the construction industry provides many job opportunities for the formal and informal sectors [1]. Any negative impact in the sector has severe consequences on the economy. One common phenomenon bedevilling the attainment of sustainable development goals in Nigeria is the issue of building collapse [2]. Nigeria has continued to experience structural failure despite the measures that have been put in place to curb this incidence [3]. Several factors have been attributed to being causative factors of these incidences. Among others, poor quality of building materials has been highlighted as playing a leading role in the collapse of buildings [4]. Some cement brands available in the open market do not meet the standard requirement, possibly due to negligence on the part of the manufacturers [5]. It is, therefore, necessary, that materials used for construction purposes are tested to ensure they meet the minimum standard requirement. Most infrastructures like bridges, offices and residential apartments in Nigeria are made of concrete because it is widely available and relatively economical construction material. Concrete is a mixture of coarse and fine aggregates held together by a binder. The most commonly used binder is the ordinary Portland cement. The main qualities of cement used for civil construction are strength and a rate of setting suitable to the demands of the work [6]. Both chemical and physical tests are carried out to assess cement quality. Some of the physical tests generally performed to determine the acceptability of cement include fineness test, consistency test, setting time test, soundness test, strength test, the heat of hydration test, and specific gravity test. This study is, therefore, geared towards investigating some of these parameters in the different brands of cement available in the Nigerian market. It will determine the granular size/particle size distribution, specific gravity and bulk density of the various cement brands in Nigeria.

2. Methodology

This work involved the determination of the fineness, specific gravity and bulk density of five brands of cement in Nigeria. A preliminary survey on the different types of cement available in Nigeria was carried out to determine their relative availability. The sample points include states within the Niger Delta region in Nigeria. These samples were labelled A-E, as shown in table 1 to make an unbiased study on their properties and variation.

Table 1. List of cement brands and their corresponding labels

Sample	Brand of cement	Classification
A	Dangote 3x (R) cement	CEM II A-L 42.5 R
B	Bua cement	CEM II A-L 42.5 N
C	Dangote 3x (N) cement	CEM II A-L 42.5 N
D	Unicem limestone cement	CEM II B-L 32.5 R
E	Elephant superset cement	CEM II A-L 42.5 N

Fineness

The specific surface method was used to determine the particle sizes. This method involves using a cello-tape whose length and width was first weighted to get the weight of tape only after which it was placed on a clean flat slab. The cement sample was poured over the full length, and a form was used to spread the sample across the range of the tape until every part of the sticky side of the cello-tape was covered, then it was picked up and shook to get rid of the excess cement. This was then weighed and recorded. The specific surface was then calculated by dividing the area of the tape by the weight of the cement.

Specific gravity

The specific gravity of cement was determined using the Specific Gravity bottle according to IS: 2720-Part-3. The specific gravity was calculated using the formula:

$$sg = \frac{(w2 - w1)}{[(w4 - w1) - (w3 - w1)} \, x \, \rho l$$

Where,
W1 = weight of empty flask
W2 = weight of flask + cement
W3 = weight of flask + cement + kerosene
W4 = weight of flask + kerosene
ρ1 = density of kerosene (0.924g/cm3)

Bulk density

The bulk density was determined using a container of known definite volume which was measured using a Vernier calliper and calculated using the formula for the volume of a cylinder. The cement was poured into the cylindrical container in three layers. It was tapped after each layer to remove air voids. A knife-edge was used to level the surface. This was to ensure that the volume of the cement equals the internal volume of the cylinder. This was then weighed and recorded. This process was repeated for all the samples.

3. Results and Discussion

From the fineness test results, as shown in figure 1, it is observed that sample A had the highest fineness value of 360m2/kg, which implies a greater surface area for hydration leading to faster development of strength and a higher rate of heat evolution. Other samples had values below 280m2/kg as specified by ASTM.

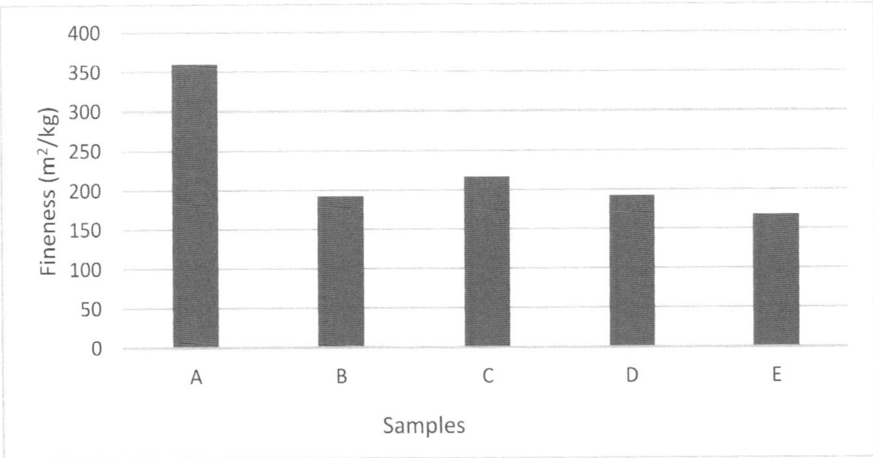

Figure. 1 fineness values for the different brand samples

Figure 2 shows the specific gravity values for the various brands' samples. Sample B had the highest specific gravity value of 3.266, implying more moisture in the sample. This has a direct effect on the workability and bonding strength of the cement as the actual water/cement ratio will be altered.

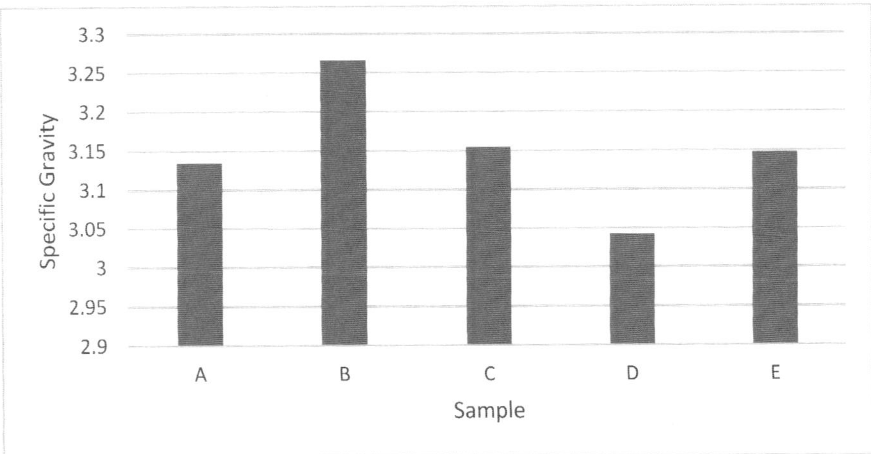

Figure. 2 specific gravity values for the different brands' samples.

Figure 3 shows that sample B had the highest bulk density value. The implication is that during mixing, a lesser volume of sample B will be required in making concrete.

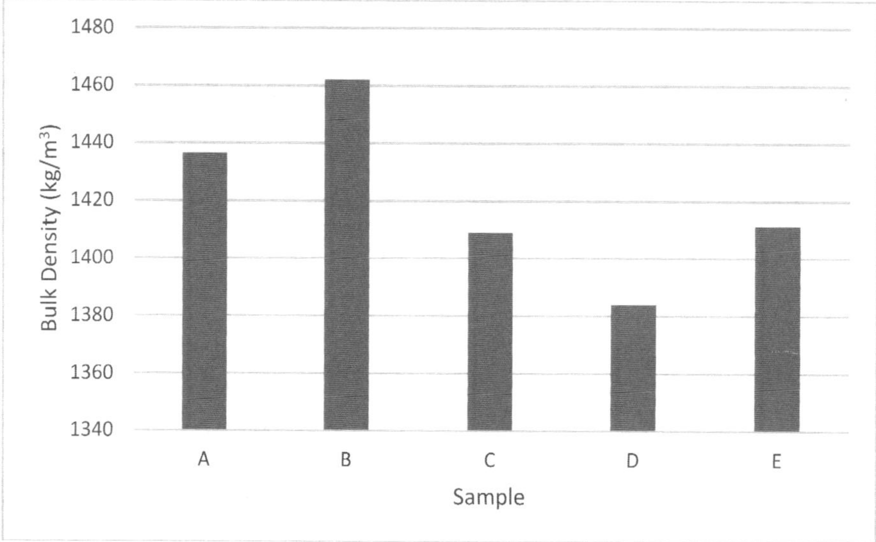

Figure. 3 bulk density values for the different brands' samples

4. Conclusion

From the results obtained, the following conclusion can be drawn.

1. Sample A has a greater surface area for hydration which should lead to faster development of early strength and a higher rate of heat evolution.
2. Sample E has the least surface area for hydration and should be slow in early strength gain.
3. Other samples except sample A did not meet the recommended minimum specific surface area value of 280m^2/kg as specified by ASTM.
4. Sample B has the highest specific gravity value of 3.266, which implies more moisture in the sample.
5. Sample B has the highest value of bulk density, which implies a reduction in sample B quantity to other samples during the concrete mix.

References

[1] Bediako M and Amankwah E O (2015) Analysis of Chemical Composition of Portland Cement in Ghana: A Key to Understand the Behavior of Cement, *Advances in Materials Science and Engineering* Vol. 2015, doi.org/10.1155/2015/349401

[2] Ede A N, Amadin I E, Akpabot,A I, Oyebisi S O, Olofinnade O M, Nduka D O, Oyeyemi K D Review of the implications of frequent building collapse in Nigeria on sustainable development goals, *International conference on sustainable infrastructural development*, Ota, 2020

[3] Akpabot A I, Ede A N, Olofinnade O M, and Bamigboye G (2019) Risks of seismic activities in built environment in Nigeria, *International Journal on Environment and Sustainable Development*, vol. 18, 259-270.

[4] Akpabot A I, Ede A N, Olofinnade O M, Bamigboye G (2018) Predicting buildings collapse due to seismic action in Lagos state. *International Journal of Engineering Research in Africa*, ISSN: 1663-4144, vol. 37, 91-102. doi: 10.4025/www.scientific.net /JERA.37.91

[5] Bamigboye G O, Ede A N, Egwuatu C, Jolayemi J, Olowu O, and Odewumi T (2015) Assessment of Compressive Strength of Concrete Produced from Different Brands of Portland Cement, *Civil and Environmental Research*, vol.7, No.8, 2015

[6] Biswal L 2017. A civil Engineer: Seven (7) Cement Test to Justify its Quality. [Online] Available at http://www.acivilengineer.com/2017/05/cement-test.html

Seismic Performance Assessment of a Typical Peruvian Public-School Building

O Cardenas[1], A Farfan[1], G Huaco[2], and A Stavridis[3]

[1]Bachelor, Peruvian University of Applied Sciences UPC, Lima Peru
[2]Professor, Peruvian University of Applied Sciences UPC, Lima Peru
[3]Associate Professor, University at Buffalo UB, New York USA

E-mail: pccighua@upc.edu.pe

Abstract. Several Peruvian public school buildings have not performed well during earthquakes, with the level of damage varying from cracks in structural elements to the complete collapse of the building. The reasons for the unsatisfactory seismic performance of the buildings include the design codes followed at the time these schools were built, and the lack of professional engineering supervision during the design and construction phases of these schools. This study focuses on a typical school building, selected after performing a survey to public schools in the district of San Juan de Miraflores in the city of Lima, in Peru. The seismic response of the selected building to various ground motions is analysed with nonlinear static and time history analyses. The results indicate that the building does not meet the minimum seismic performance level required by the current Peruvian seismic design provisions.

1. Introduction

In all areas located at highly active seismic zones, it is important to guarantee a sufficient seismic resistant behaviour of the educational infrastructure. In the 90s, the Peruvian Government built new schools throughout the country, according to the so-called Construction Plan [1]. A recent qualitative study carried out about the seismic risk of public educational infrastructure [2], indicates that the "780 Pre" building is a representative building that can be found in many cities of Peru, many of them at high seismic zones.

The "780 Pre" is a two-story building with plan dimensions of 7.80m x 23.5m and 3.35m height per story [3]. It was built before 1997, typically not following current design and construction practice [2]. Moreover, after earthquakes events serious structural damage has been observed in this type of building, including shear failures on columns due to the short column effect introduced by the infill walls and the short wide windows at the top introduced to allow light into the classrooms [4]. The short column effect and typical distribution of the "780 Pre" building are shown in Figure 1. In this article, the seismic performance of such buildings is investigated with static and dynamic non-linear analyses.

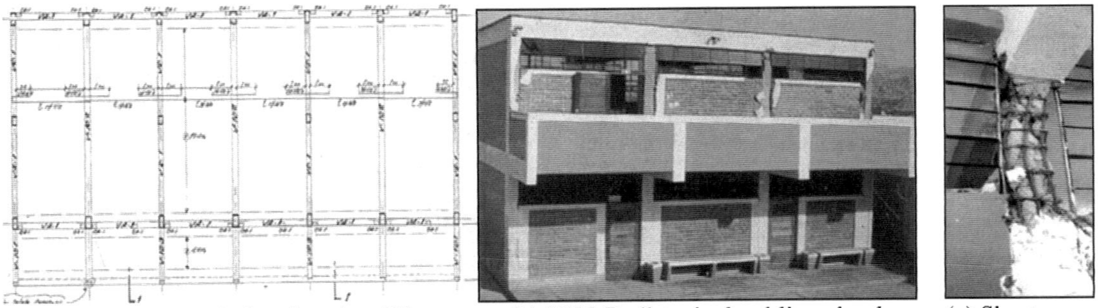

(a) Original plan drawing [3]

(b) "780 Pre" typical public-school building after the 1996 earthquake [1]

(c) Short column effect due to 2011 earthquake [2]

Figure. 1 a), (b), and (c)

2. Analysis Procedures

2.1. Computational Model

The building consists in confined masonry in the transverse axis and reinforced concrete frames with masonry infills in the longitudinal axis. Concrete column sizes are 25cm x 45cm with 6Ø¾" bars, while the masonry walls are 25cm thick when parallel to the transverse direction of the building and 15 cm thick when aligned with the longitudinal direction [3]. It is modeled using the software SAP2000. In the transverse direction, the masonry walls were modelled as equivalent struts as proposed by Bazan & Meli [5]. Material strength for concrete and masonry in compression, and steel in tension are f'c of 17MPa, f'm of 4.4MPa, and fy 420MPa, according to original drawings [3].

In the longitudinal axis, infill masonry walls were initially modelled as shell elements with elastic behaviour because there was no significant damage seen on these walls after previous seismic events. These infill walls restrain the lateral displacements of columns up to a certain height, causing excessive shear stresses in the remaining free height. For comparison purposes, the infill masonry walls are also modelled as equivalent struts using the same methodology as the transverse masonry walls. Figure 2 shows the type of models per case.

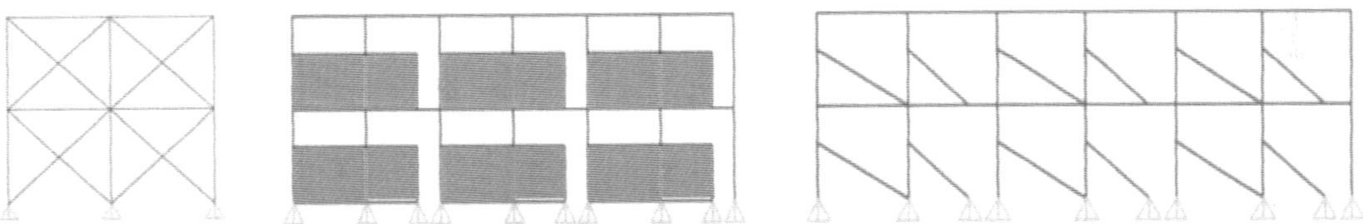

(a) (b) (c)

Figure. 2 Transverse axis masonry walls modelled as equivalent struts(a), infill walls modelled as shell elements (b) and equivalent struts (c)

Plastic hinges are defined at the ends of beams and columns For concrete columns the moment hinge represents a flexure-controlled failure of the element, meaning that the reinforcement steel, concrete strength or element dimensions are not sufficient to resist the moments that are developed in the location of those hinges. The shear hinge is used to capture the short column failure observed in these schools after different seismic events [4]. For masonry elements an axial hinge is applied in the equivalent struts and represents the failure in tension/compression of the masonry and the infill walls, in the transverse and longitudinal directions, respectively.

2.2. Capacity Curves from Non Linear Static Procedures NLSP

A monotonically increasing base shear force is applied to the structure until certain displacement in a control joint is reached. As result, capacity curves are obtained per each orthogonal direction and they are shown in Figures 3 and 4.

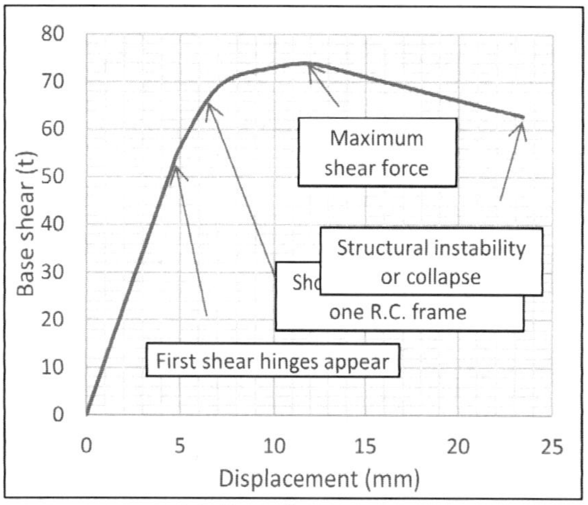

(a) Capacity curve

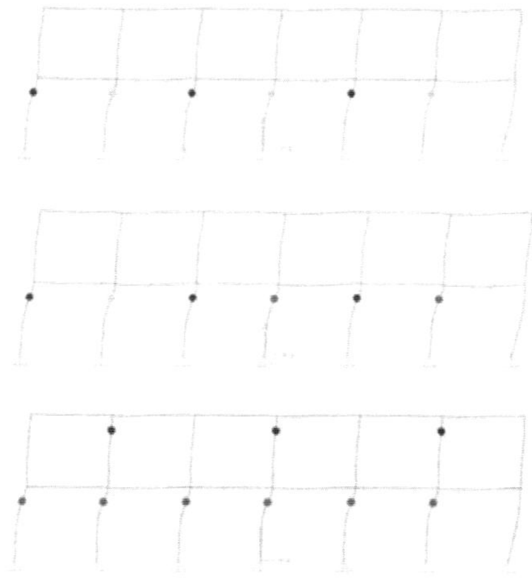

(b) Shear hinges

Figure. 3 Pushover analysis results along the longitudinal direction

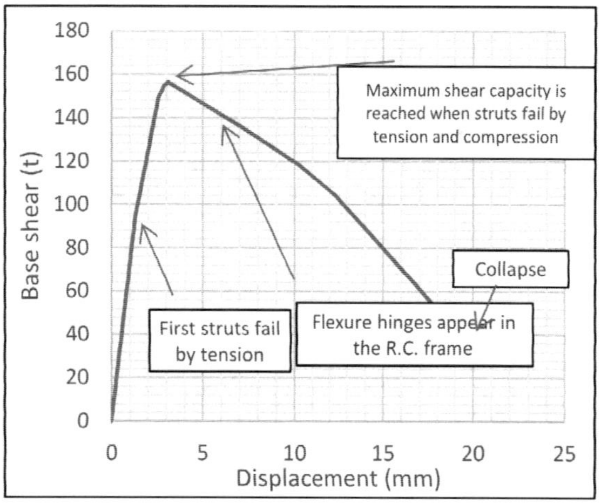

(a) Capacity curve

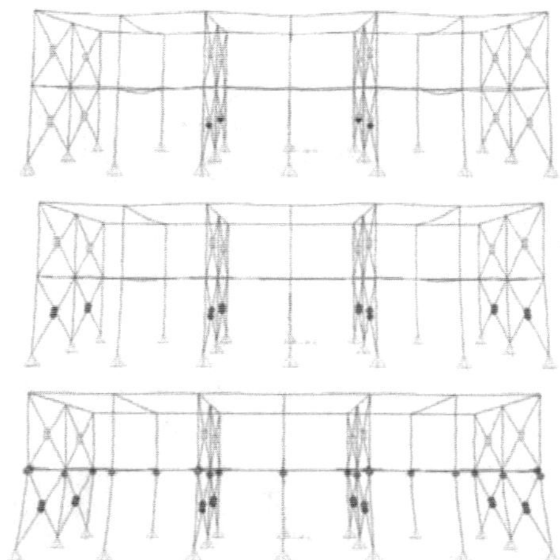

(b) Damage pattern

Figure. 4 Pushover analysis results along the transverse direction

2.3. Capacity Curves from Non Linear Time-History Analysis NLTHA

Non Linear Time-History Analysis (NLTHA) is used to assess the behaviour of the structure when it is subjected to cyclic loads due to ground motion records of a real-life earthquake of 6ML magnitude [6].

Given that the longitudinal axis is the structural weaker direction according to the typical collapse mechanism seen in these buildings in previous seismic events, it is the only axis that is analyzed through the NLTHA. The envelopes in Figure 5 represent the capacity curves of the structure modelled with different approaches, including shell elements and diagonal struts for the infill walls.

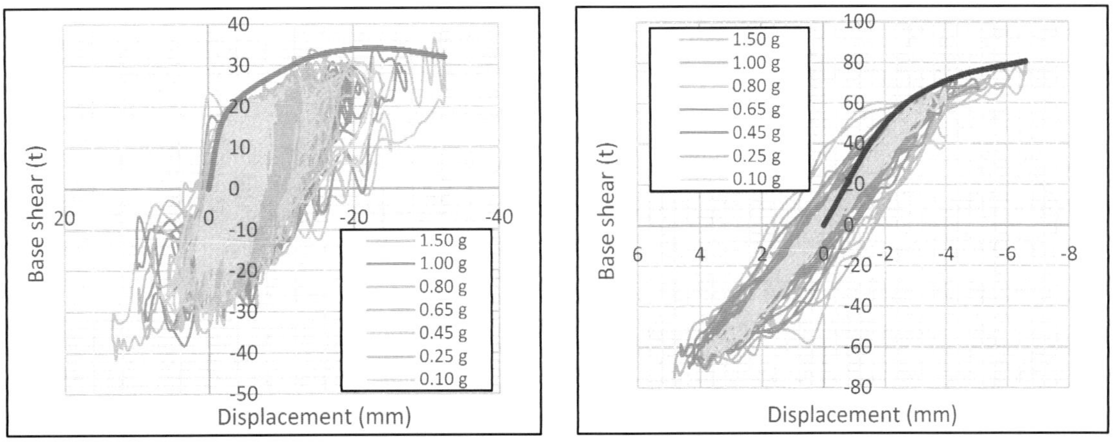

(a) (b)

Figure. 5 NLTHA results for longitudinal direction using equivalent struts (a), and shell elements (b)

2.4. Performance Based Seismic Design

After performing the non-linear analyses, the performance point assessment is found. This method is proposed by the ATC-40 [7], therefore the ADSR conversion equations were used.

Four different soil conditions from the Peruvian seismic design provisions [8] are assessed in this study: S0, S1, S2 and S3with S0 soil representing a stiff rock soil and the S3 a flexible weak soil. The spectrum responses according to different soil conditions are converted to seismic demand curves. Then, these curves are plotted with the capacity curves found in the previous section for both elastic and inelastic demand curves. The performance levels are also plotted along the other curves, as shown in Figures 6 and Figure 7.

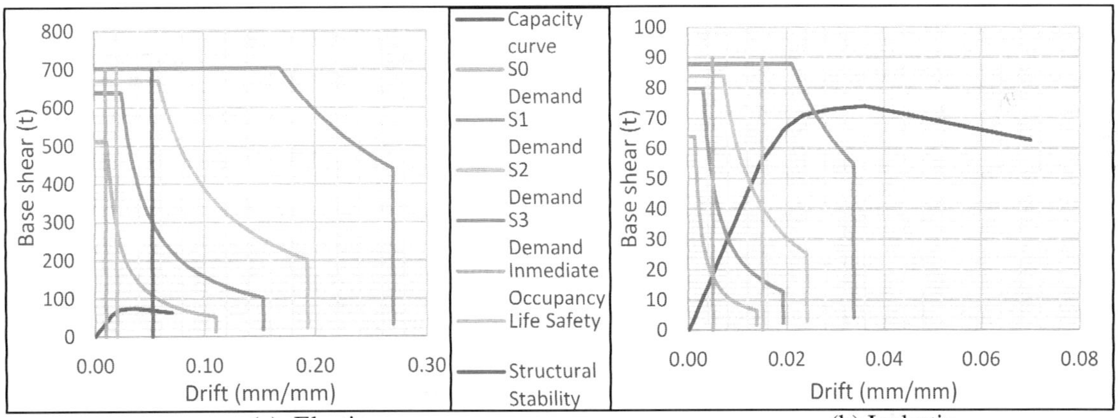

(a) Elastic (b) Inelastic

Figure. 6 Seismic performance assessment for the longitudinal direction with moment frames

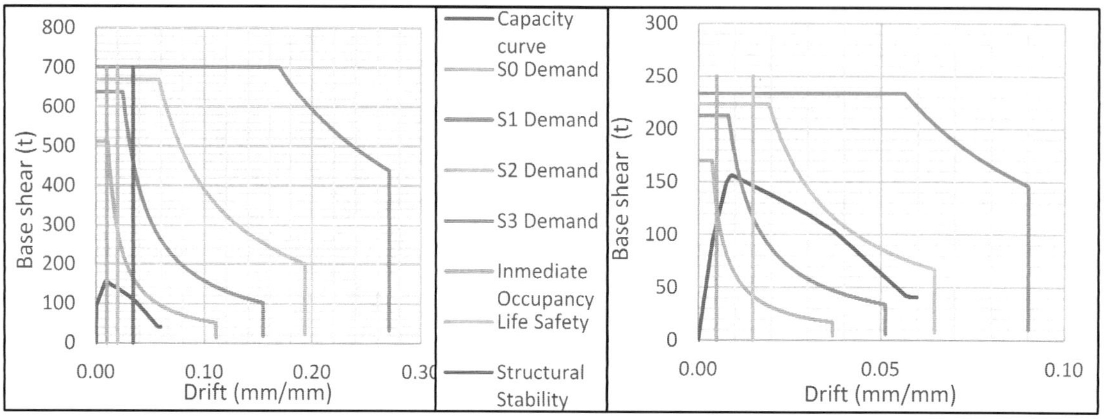

(a) Elastic (b) Inelastic

Figure. 7 Seismic performance assessment for the transverse direction with confined masonry walls

3. Discussion

The NLSP in the longitudinal direction indicates that the structure fails by excessive shear stresses in the columns caused by the infill masonry walls. The highest base shear force that the structure is able to resist in the longitudinal direction is 73.9t with a lateral drift of 1.79‰, while the drift before collapse is 3.43‰.

In the transverse direction, the failure mechanism initiates with the cracking of the masonry walls, which end up failing in compression. The points after the peak of the capacity curve represent the influence of the reinforced concrete frame that is supposed to only serve as confinement of the masonry wall. At the end, the structure collapses due to the excessive moments in the plastic hinges defined in the frame elements of the reinforced concrete confinement frame. The capacity curve in this direction indicates that the structure can resist a base shear force of 156t associated to a 0.45‰ drift. The structure would collapse in this direction if the lateral drift reaches 2.99‰ as the resistance drops to 41t.

With the NLTHA, two capacity curves are obtained by drawing the envelopes over the hysteresis curves of both models, with shell elements and equivalent struts. These capacity curves are also compared in one graphic with the capacity curves from NLSP results, for the two modelling variations, as shown in Figure 8. The difference between these curves shows that the model is sensitive when the infill walls are modelled with an elastic or inelastic response, and when there is a monotonic NLSP or dynamic NLTHA load analysis, as expected.

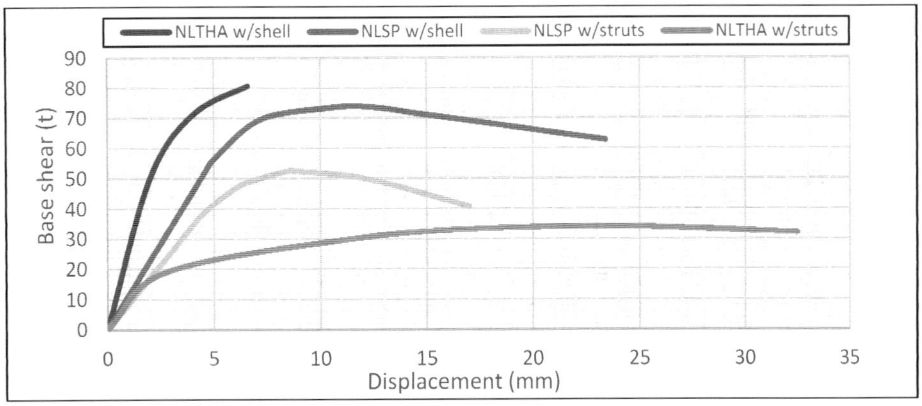

Figure 8. Capacity curves for the longitudinal direction

The seismic performance assessment of the longitudinal direction shows that when the structure is expected to behave elastically, it is not able to intercept any of the mentioned curves. This means that it would not be able to resist the base shear force even if it were built upon a S0 soil, the best soil considered in this study. On the other hand, when the structure is subjected to inelastic seismic demands, it does intercept these curves. Hence, the structure's performance point for the S0 soil is immediate occupancy, for the S1 and S2 soils is damage control, and for the S3 soil it is beyond the damage control limit.

In the transverse direction, it is found that the structure does not intercept any of the elastic seismic demand curves, indicating that it would not resist the base shear forces if it were built upon any soil condition. Compared to the inelastic demand curves, the structure only intercepts the S0 and S1 soil. For the S0 soil, it still has capacity to resist slightly higher seismic accelerations, but for the S1 soil the structure reaches its inelastic behavior. For both S0 and S1 soils, the seismic performance level is of damage control.

4. Summary and Conclusions

Both static and dynamic non-linear analyses are performed to the "780 Pre" typical public-school building of Peru, using different approaches when modelling the infill masonry walls. The seismic behaviour is also assessed by plotting its capacity curve in each direction and comparing against the spectrum responses estimated for the four different soil conditions, which were converted in compatible plots of base shear force and lateral drifts.

It is found that the typical school building does not guarantee its immediate occupancy use as a refuge infrastructure after a severe earthquake in most cases. The only exception is if the structure is built upon at a site with S0 soil with a seismic acceleration in the longitudinal direction.

Analyzing effective methods for the structural strengthening of these buildings is highly recommended for all "780 Pre" buildings found throughout the country.

References

[1] Blanco J 2005. *Twentieth-Century Educational Structures in Peru, Diverse Strengthening Projects and Examples from Catholic University of Peru Buildings* (Lima) pp 17-21

[2] O. Cardenas, A. Farfan and G. Huaco, *Seismic Risk Assessment of Peruvian Public School Buildings Using FEMA P-154 Rapid Visual Screening.* 2020 Congreso Internacional de Innovación y Tendencias en Ingeniería (CONIITI), Bogotá, Colombia, 2020, pp. 1-5, DOI: 10.1109/CONIITI51147.2020.9240369.

[3] National Institute for Infratructure of Education and Health INFES 1993, Construction Drawings of 780 Pre for school buildings, Peru

[4] Elnashai, A., Alva-Hurtado, J., Pineda, O., Kwon, O. S., Moran-Yanez, L., Huaco, G., & Pluta, G. (2008). The Pisco-Chincha Earthquake of August 15 , 2007. No. 08-01

http://hdl.handle.net/2142/9460
[5] Bazan E and Meli R 2002 *Seismic Design of Buildings* (Mexico), Mexico
[6] Acelerographic Network, National University of Engineering – Graduate School Civil Engineering Faculty & Peru Board of Engineers, Posgrado UNI-FIC & CIP Palpa Earthquake 2019
[7] Applied Technology Council 1996 *Seismic evaluation and retrofit of concrete buildings Report No. SSC 96-01: ATC-40*
[8] Peru Construction Codes – Seismic Design Code E.030, 2018. National Institution Of Guidance To Construction Industry. SENCICO, Peru

A novel visual inspection system for rail surface spalling detection

Dieuthuy Pham [1,2], Minhtuan Ha [1,2] and Changyan Xiao [1]
[1] College of Electrical and Information Engineering, Hunan University, Changsha 410208, China
[2] Department of Electrical Engineering, Saodo University, Hai Duong 170000, Vietnam

Abstract. Railway transportation plays an important role in economic development since it can transport large volumes of passengers and cargo through great distances. Therefore, monitoring the condition of the railroad is essential to ensure train safety. With the development of computer vision, the railway non-destructive inspection systems have become possible. However, these methods have been still challenged by so many obstacles from ambient light on the rail surface and defects themselves. Defects appear on the rail surface are variety, in which spalling type is usually in heterogeneous shape and size. In this paper, a visual inspection system for rail surface spalling detection is proposed. The track image is first segmented by a novel rail track extractor. Then the rail surface spalling can be coarsely detected based on histogram curves in the longitudinal direction of the track image. Finally, a fusion technique is performed to eliminate all the false detected defects in the resulting image. Experimental results demonstrate that the proposed method achieves the precision and recall of 97.48% and 95.74%, respectively, and shows good robustness under nonuniform illumination and various rail surface conditions.

1. Introduction

As abovementioned, the railway inspection is essential to ensure train safety. Traditionally, one well-trained person has to walk along the railroad to evaluate the rail condition. However, this kind of inspection has many crucial disadvantages such as waste of human resources, time-consuming, subjective evaluation, and the performance of a worker may be degraded within one day and serious mistakes can be made. Thus, automatic inspection systems are required for an objective, fast, and reliable quality assessment of railways.

To handle this goal, a lot of paper that differ in either inspected objects or approach methods have been developed, including Fiber Bragg Grating sensors [1-3], ultrasonic inspection [4], eddy current [5], MEMS gyroscope sensors [6], and visual inspection system (VIS) [7]. Especially, VIS with its advantages has been developing and become more and more accurate. For instance, Mandriota et al. [8] performed three filtering approaches based on texture analysis and chose the Gabor texture feature for rail surface images, and then apply the k-nearest neighbor (KNN) method to identify the possible defects. In [9], the geometrical features (i.e., area, centroid, filled area, and perimeter) of the defected regions based on maximally stable extremal region technique are extracted. Li et al. [10] proposed the method of LN+DLBP for real-time discrete defects inspection using the local normalization (LN) method, which is nonlinear and illumination independent then applied the defect localization based on projection profile (DLBP) algorithm to identify possible defects. In [11], a new inverse Perona-Malik (PM) diffusion model is presented for image enhancement and, subsequently, an adaptive threshold binarization can easily find out the desired defects. Although these approaches have achieved certain effectiveness in detecting several types of defects none of them aim to inspect spalling with extreme various appearances and dynamic backgrounds due to heavy random noise caused by complex rail track conditions and camera quality.

This paper presents two scientific contributions: an accurate rail track extractor and rail surface spalling detection method for the railway inspection system.

Accurate extraction of ray surface image plays an important role in detecting defects on the rail surface. However, this problem has not been completely solved by the previous studies, so it is only possible to extract the rail images under certain assumptions. In this study, we carefully analyze rail imaging conditions such as railroad background, ambient light, and contrast in each acquired image, therefore it is easy to accurately extract rail surface image.

To overcome the challenges of detecting spalling on the rail surface, our approach takes two main steps:

1) Determine longitudinal defect according to the maximum gray value in each column of rail image individually.

2) Combine with a complete adaptive threshold image for eliminating false defects in longitudinal consideration.

The remains of this paper are organized as follows. After our system is introduced in Section 2, an accurate rail track extractor is presented in Section 3. Then Section 4 describes a method for detecting spalling on the rail surface. The experiment results are shown in Section 5, followed by our conclusions in Section 6.

2. System overview

This paper proposes a real-time visual inspection system (VIS) for spalling detection on the rail surface and produces an accurate rail track extractor. VIS consists of the image acquisition subsystem (IAS) and the image processing subsystem (IPS). The basic component of IAS is a Dalsa Spyder 2 line-scan camera with a resolution of 1024 x 1024 pixels and a maximum line rate of 65 000 lines/s. A PC-CamLink frame grabber is utilized to capture rail images based on the camera link protocol. An illumination setup equipped with two LED light sources is installed to reduce the effect of ambient light. Moreover, the line-scan camera is triggered by a wheel encoder to synchronize data acquisition.

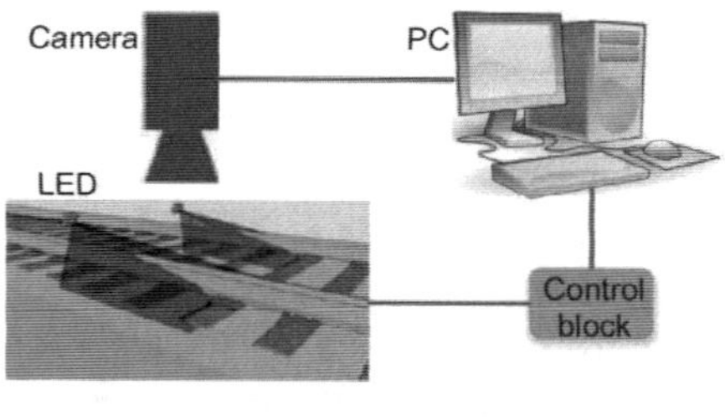

Figure. 1 Our system configuration

Our system configuration (Figure 1) has some advantages as follows:

+ Reflection property of rail surfaces in the longitudinal direction is pretty homogeneous.

+ An auspicious installation to detect rail track longitudinal defects.

+ Easy for the operator to install and adjust.

Our system installation specification also remains some disadvantages, such as some small bright or dark regions probably appear on the rail surface image or the shadow of rail track exists in the acquired image. However, the proposed method overcomes these obstacles easily. For example, the solid dark region evaluating can be applied on both sides of the rail image for finding the rail shadow. And since the brightness intensity is homogeneous on the rail surface in the rail track longitudinal orientation, thus the abnormal solid bright regions (shown in Figure 6g bottom right) would be eliminated by performing an image fusion ceremony of defects detected rail images in both longitudinal and transversal direction.

3. Rail track extractor

In the past few years, various texture analysis approaches have been proposed, which are widely applied in many applications such as classification, inspection, and segmentation of image based on the histogram gray value curve (Rail localization based on Projection Profile (RLPP) [10], Rail Localization based on Weighted projection Profile (RLWP) [12], Geometrical defect Locating method based on Gray-level Histogram Curve (HGLHC) [13]) of the original image or Hough transformation. These methods only perform well with some certain assumptions of the rail image, such as the maximum deviation of the gray value of the pixels in each column in the rail image is not too large, and the rail surface region is the brightest. However, as we can see in Figure 2, such assumptions are often broken by illumination inequality and the variation of reflection property of rail surfaces and ballast or concrete background.

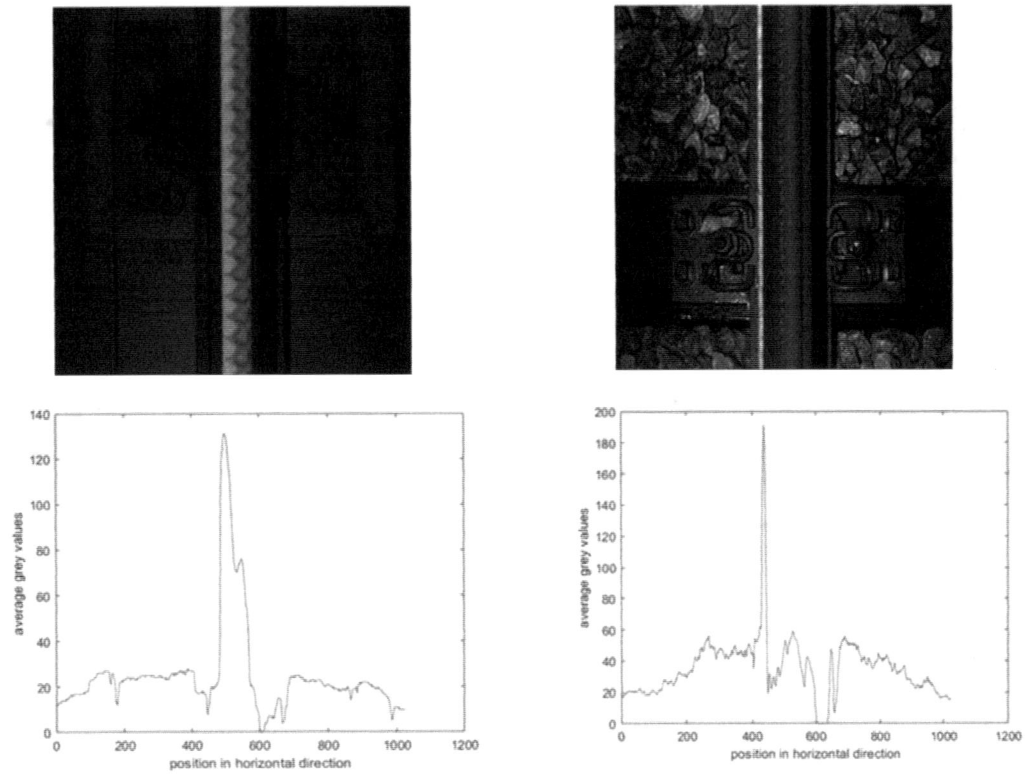

(a) (b)

Figure. 2 Two samples of rail image in our data set. (a) Rail image with a concrete background. (b) Rail image with ballast background

On the other hand, some disadvantage conditions challenge the methods based on Hough transformation such as lots of strikes or lines that appeared on the original image beside the rail track (Figure 3a). Moreover, it is difficult to exactly localize the rail track region if serious defects exit on the rail surface or rail track shadow cast alongside the rail track because their features are always variable throughout the images. To decrease the impact of these, a series of transformations need to apply, but they complicate the algorithm and cost processing time.

As we all know, rail images obtained in reality are influenced by the imaging conditions as well as the quality of the rail surface. So if just relying on some assumptions like such as the gray value of the pixels on the rail surface is relatively homogeneous and larger than the gray value of other areas in the image or the rail track is in the middle position of the image obtained from the imaging system ... will be unable to accurately extract the rail track images. To overcome these obstacles, we perform the following steps:

1) Identify the background of the rail image (ballast or concrete) based on the average gray value deviations in each column of the whole image. In our experiments, the background is ballast as that value is larger than 1500.

2) The shadow of a rail track in the image always has the nearly smallest gray value in the whole image. Therefore, it can be determined as the largest region in the longitudinal direction of the binary image with a small threshold value (Figure 3b).

3) By accurately localizing the shadow of the rail track, the image of the rail surface will be determined by comparing the two adjacent areas of the shadow. If the background is a ballast, the rail track image will be the area with the larger average gray value deviations in each column (the gray value of the pixels in the rail surface region is more uniform than the ones in the background region). If the background is concrete, the rail track image will be the region with the larger average gray value in each column (rail surface is brighter than the concrete surface).

4) Use a rectangle with exactly rail track dimensions to extract the rail track image.

(a) (b) (c)

Figure. 3 Illustration of rail track extracting procedure. (a) Original image. (b) Binary image with a small threshold. (c) Extracted rail track image.

4. Rail surface spalling detection

Defects are easy to be hidden or confused in rail images because of illumination inequality and the variation of reflection property of rail surfaces, so contrast enhancement is a necessary procedure to highlight defects from their background. Different from [14], the proposed method aims to detect spalling on the rail surface image of which gray value is larger than adjacent regions in the longitudinal and transversal direction respectively. Some of the main reasons are as follows:

- Defect's features (shape, size, gray value) always change from image to image and challenge the visual inspection systems. The defect region may be very small with a size less than 20 pixels or even very large with a size greater than 10000 pixels, which spread though over the image in the longitudinal direction. Consequently, it is unreliable to evaluate the maximum deviation of the gray value in the longitudinal direction because some large defects could be omitted (Figure 4).

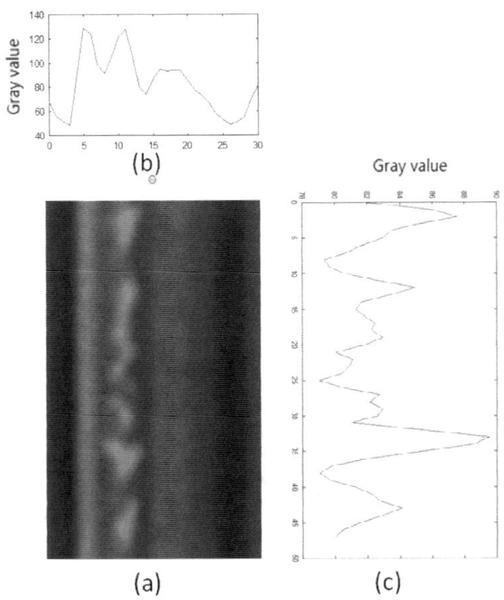

Figure. 4 Gray value Histogram curvature of a rail sample image (a) in longitudinal (b) and transversal direction (c)

- Generally, the spalling is brighter than the background. However, this order can be changed because of variations of illumination and reflection property or some small lighter or darker regions.
- Furthermore, numerous objective noises caused by visual inspection systems establishing and operating processes significantly affect the quality and performance of acquired images. The noticeable ones are time, weather, place of inspection, natural light, quality of the image acquisition subsystem (IAS), the velocity, and the shake of the moving test train. All of them need to be particularly considered for cancellation in pre-processing.

According to the above analysis, the proposed method is based on the maximum gray value deviation in each column individually then associated with a complete adaptive threshold image in the result.

4.1. Pre-process

Firstly, the mean gray value of the whole rail image is calculated. After that, all the small regions which have too large a gray value while comparing with the mean gray value of the whole image are eliminated.

According to the mean gray value, the contrast of the rail image is adjusted. Then a boundary detecting method by evaluating the Gradient of pixels is performed.

Since the digital image is discrete signals so the partial differential equations are replaced by performing a convolution technique to get the linear approximation result.

Following the above analysis, a method of which advantage to eliminate the impact of random noises and keep the image detail simultaneously is applied (Equation (1)):

$$R(x,y) = \alpha * f(x,y) + (1-\alpha) * g(x,y) \quad (1)$$

where x and y are the spatial coordinates of acquired raw image *f(x,y)*, *g(x,y)* represents the smoothed image of *f(x,y)* by a median filter, *R(x,y)* denotes the output image. And α(x,y) is the smoothing rate parameter which can be derived from Equation (2).

$$\alpha(x,y) = \frac{\partial g(x,y)}{\partial x} / \max\left(\frac{\partial g(x,y)}{\partial x}\right) \quad (2)$$

4.2. Rail surface spalling detection method

As shown in Figure 5, the process of spalling detection is composed of the following main steps:

1) In each column of the image, find out the maximum and minimum gray value pixels, and calculate the mean gray value of the pixels in that column simultaneously. If the maximum deviation of the pixels in that column is larger than a threshold value then there are defects in that column. After that, the

contrast of these columns in the image is adjusted by performing a quadratic function $y = k*x^2$, where $k = 0.01$. Meanwhile, all the pixels in other columns of the image are set to zero. Finally, an adaptive threshold is applied to the retrieved image to obtain a binary image with defects detected in the longitudinal direction.

2) Another binary image obtained by performing an adaptive threshold on the original rail image.

3) Combine the above two result images using a bit AND logical function to eliminate non-defect regions on both of them and reshape the derived defects simultaneously.

By performing a connected component detecting function with the 8-nearest neighbor parameter, the size of each defect can be determined. In accordance with that, exactly identify all the defects and their features, thenceforth evaluating the harmful of them to railway operating will be available.

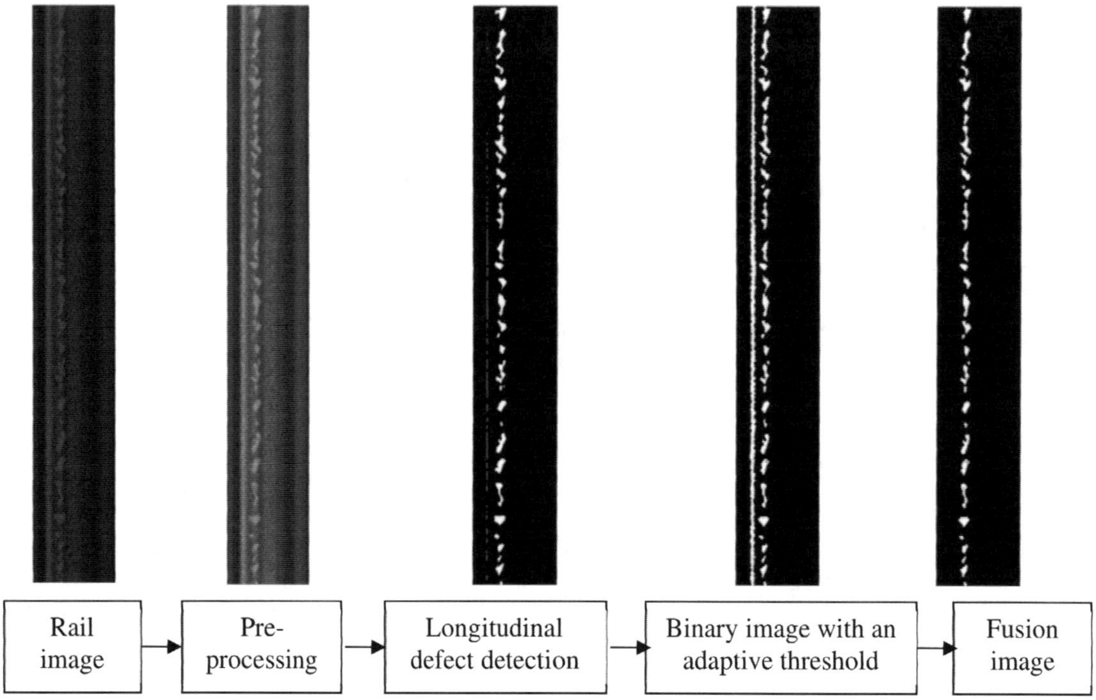

Figure. 5 Flowchart of rail surface spalling detection process.

5. Experimental results and discussions

Our experiments are conducted with the images collected in the actual railway of China by the abovementioned IAS and divided into two small data sets. One including 500 images with several abnormal illuminations and different types of defects on the rail surface which are arbitrarily chosen from a large image set is used for evaluation of rail track extractor. And the other contains 1108 images which consist of 556 typical spalling rails and 552 normal rails.

5.1 Performance evaluation of rail track extractor

In our experiments, the ray width in the images is a constant of 120 pixels. The difference of this method compared to the previous ones is that it ignores some assumptions such as the rail surface is brighter than the background, the gray level value of the pixels on the rail surface area is relatively homogeneous or the position of the rail track is at the center of the image. Our experiments show that those assumptions are broken easily for several reasons. For example, the vibration of the test train while running in curves makes the position of the rail track surely deviate from the center axis of the acquired image. For images with many large defects, the surface of the rails is so deformed that resulting in gray values of the pixels on the rail surface area in the longitudinal direction becomes inhomogeneous. Also, ambient light noise such as contrast and brightness of the background can affect the image quality and make the rail surface region no longer brighter than the others.

It can be seen that the rail shadow is a homogeneous region in which the pixels have the smallest gray value in the image. Therefore, using our method, it can be detected easily based on the average gray curve in Figure 2. After that, the rail track image can be localized by comparing the two regions of which the width is equal to the width of the rail lying alongside the shadow. In that way, the proposed method can extract the rail track image at any position in the original image. Quantity statistics of experimental results show that the performance of the proposed rail track extractor reaches a value of 98.6%.

5.2 Performance evaluation of spalling detection method

As we can see in Figure 6, spalling appears on the rail surface in various shapes, colors, and sizes. They may be very small, scattered on the rail surface, but can also be large and spread along the entire length of the rail image. However, many traditional methods fail to detect the spalling of such large sizes by mistaking them for light noise on the rail surface and ignoring them.

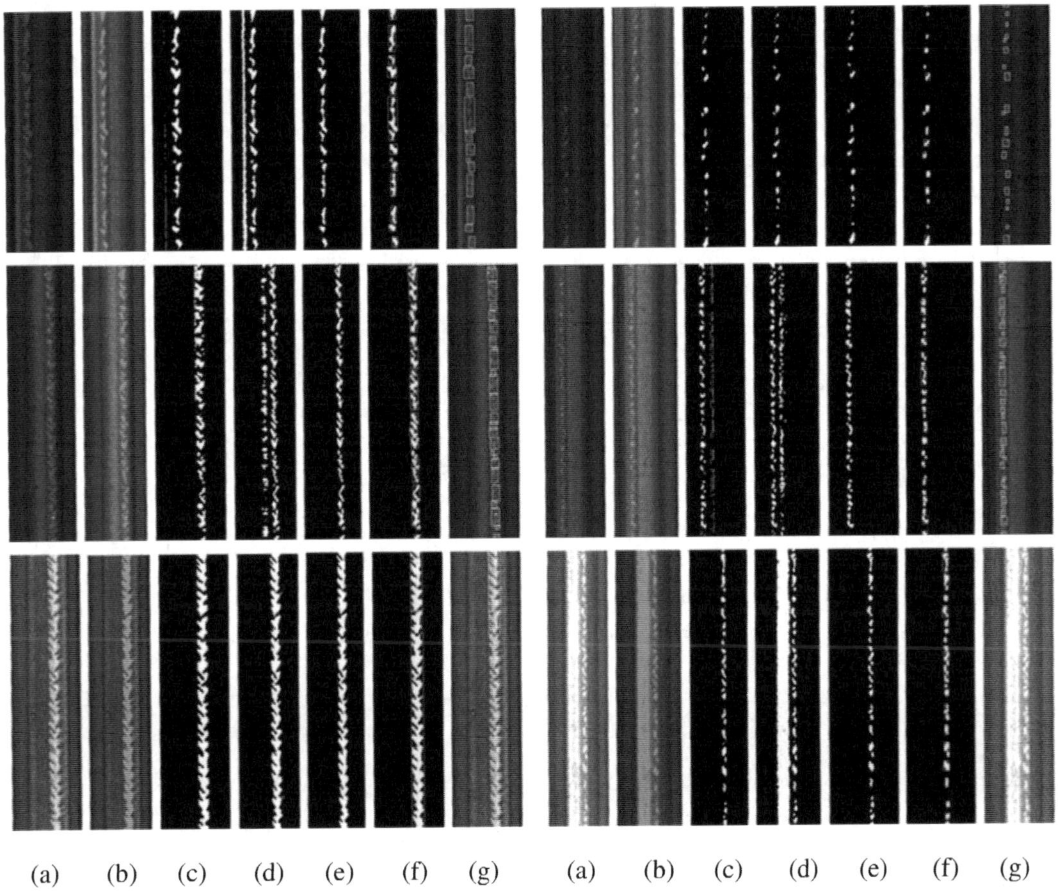

Figure. 6 Some samples and the results derived with our method. (a) Original rail image; (b) Image after pre-processing. (c) Longitudinal contrast adjusting. (d) Binary image with an adaptive threshold. (e) Fusion image. (f) Connected components localization. (g) Spalling detection.

With the proposed method, every location with a large gray level variation in each column of the rail image can be a suspect area. Then the contrast in the corresponding columns will be raised while the gray level of the pixels in the remaining columns of the rail image are set to 0. From that, defects can be coarsely detected in the longitudinal direction of the rail image. Some defects that might be false detection in the image with defects detected in the longitudinal direction can be eliminated by combining this image with a binary image obtained from the original image with an adaptive threshold.

To quantitatively evaluate the proposed method, we use criteria of precision (pre) and recall (rec) which are defined as equations (3) and (4):

$$\text{pre} = \text{TP}/(\text{TP} + \text{FP}), \quad (3)$$

$$\text{rec} = \text{TP}/(\text{TP} + \text{FN}), \quad (4)$$

where TP indicates the number of correctly detected spalling images, FN means the number of missed spalling images, and FP refers to the number of misjudged spalling images.

Experimental results demonstrate that the proposed method achieves the precision and recall of 97.48% and 95.74%, respectively, and shows good robustness under nonuniform illumination and various rail surface conditions.

6. Conclusion

This paper has proposed a method for detecting spalling appearing on rail surface mainly based on the histogram gray level curvature in both longitudinal and transversal direction. The results of experiments conducted on 1108 rail images shown that the proposed method could easily overcome the challenges such as ambient light inequality, the vibration of the testing-car while moving, the inhomogeneous reflection property of the rail surface, defect distribution on rail images, ... Since then, the proposed method has been demonstrated to be robust to noise and efficient performance, and consequently be conveniently applied to the railway visual inspection system. Besides, this paper also produces an accurate rail track extractor with better efficiency in reality applications comparing with the conventional methods. With the detection method for the spalling defect on a rail surface, relevant railroad period maintenance planning would be made. This result also contributes a significant data source for the complete railway visual inspection system based on deep learning in future work.

Reference

[1] Filograno M L 2012 Real-time monitoring of railway traffic using fiber Bragg grating sensors *IEEE Sensors Journal* **12** 85-92
[2] Wei C L 2010 A fiber Bragg grating sensor system for train axle counting *IEEE Sensors Journal* **10** 1905–1912
[3] Filograno M L, Corredera P, Rodríguez-Plaza M, Andrés-Alguacil A, and González-Herráez M 2013 Wheel flat detection in high-speed railway systems using fiber Bragg gratings *IEEE Sensors Journal* **13** 4808–4816
[4] Li Y, Trinh H, Haas N, Otto C and Pankanti S 2014 Rail component detection, optimization, and assessment for automatic rail track inspection *IEEE Transactions on Intelligent Transportation Systems* **15** 760-770
[5] Gao B, Bai L Woo W L, Tian G Y, and Cheng Y 2014 Automatic defect identification of eddy current pulsed thermography using single channel blind source separation *IEEE Transactions on Instrumentation and Measurement* **63** 913–922
[6] Broquetas A 2012 Track detection in railway sidings based on MEMS gyroscope sensors *Sensors* **12** 16228-16249
[7] Li Q and Ren S 2012 A visual detection system for rail surface defects *IEEE Transactions on Systems, Man, and Cybernetics, Part C (Applications and Reviews)* **42** 1531-1542
[8] Mandriota C, Nitti M, Ancona N, Stella E and Distante A 2004 Filter based feature selection for rail defect detection *Machine Vision and Applications* **15** 179–185
[9] Dubey A K and Jaffery Z A 2016 Maximally stable extremal region marking-based railway track surface defect sensing *IEEE Sensors Journal* **16** 9047-9052
[10] Li Q and Ren S 2012 A real-time visual inspection system for discrete surface defects of rail heads *IEEE Transactions on Instrumentation and Measurement* **61** 2189-2199
[11] He Z, Wang Y, Yin F and Liu J 2016 Surface defect detection for high speed rails using an inverse PM diffusion model *Sensor Review* **36** 86–97

[12] Li Q, Shi Z, Zhang H, Tan Y, Ren S, Dai P and Li W 2018 A cyber-enabled visual inspection system for rail corrugation *Future Generation Computer Systems* **79** 374-382
[13] Lin J, Luo S, Li Q, Zhang H and Ren S 2009 Real-time rail head surface defect detection: A geometrical approach *IEEE International Symposium on Industrial Electronics* **Seoul** 769-774
[14] Hu Z, Zhu H, Hu M and Ma Y 2018 Rail surface spalling detection based on visual saliency *IEEJ Transactions on Electrical and Electronic Engineering* **13** 505-509

Fire Characteristics of Light Steel Structure Building and Suggestions for Fire Alarm System

Yu-Chih Ou[1], Ming-Lun Tsai[1], Hao-Yu Wu[1], Chung-Chyi Chou[1,2,*]

Department of Environmental Engineering, Da-Yeh University[1,2]
Department of Fire and Safety, Da-Yeh University[2]

dustin@mail.dyu.edu.tw

Abstract. This study is a small residential factory with a lightweight steel structure for fire characteristics analysis and alarm equipment analysis. This kind of factory fire grows very quickly, and there is no perfect fire protection zone, the temperature rises rapidly when the fire occurs, and it is difficult for the fire brigade to rescue. Use FDS for fire simulation and discuss the activation time of fire alarms with different functions, when the people in the building are awake, the thermocouple distribution position is combined with the heating curve to analyze the critical value of the human body's dangerous temperature, and the building is partitioned and the allowable evacuation time includes: 13 minutes in the factory area, 11 minutes in the corridor, and office area there are 12 minutes and the room has 15 minutes. In addition, the difference in the action time of the two detectors, the smoke detection type and the fixed temperature type, is discussed. In the simulation results, the smoke alarm device can detect the fire 137 seconds earlier.

1. Introduction

Lightweight steel structure buildings are a type of building that can be seen everywhere in Taiwan. Because of their simple structure and rapid construction, they become factories, warehouses or general temporary buildings. The common forms are own factories, roof tops, and temporary factories. (Chen 2001), this type of building has a higher risk of fire, because the structure is not strong and the temperature is likely to rise quickly during fire; the fire grows quickly; there is no building fire protection inside. (Chen 2019), the human body is exposed to high temperature and heat during a fire. When the average temperature reaches 200℃, serious injuries will occur after five minutes. About the evacuation time of personnel put forward the content of Formula 1 according to the sequence of fire and refuge action, using the concept of time.

$$\frac{Tp+Tr+Ta}{Tf} \leq 1 \qquad (1)$$

This research simulates a lightweight steel structure factory building, and uses the FDS fire simulation program to conduct fire simulation and disaster prevention research. Fire simulation items include changes in the internal temperature of the iron sheet, comparison of detector types and operating time, fire heat release rate, etc.

In the fire alarm equipment, the driving principle of the fire detector is divided into: temperature, ion type, photoelectric type, etc. (Peng 2018) Temperature detector: When the surrounding temperature rises above a certain temperature, it will immediately sound an alarm. Smoke Detector: After the smoke concentration increases, an alarm will be issued immediately when the smoke concentration in the

detection protection area reaches a certain level. The biggest factor affecting the opening of the building caused by fire is glass and combustible materials. Skelly (1991) measured the temperature and time of glass rupture in a fire environment by experiment. According to the experimental results, it is easier to break when the edge of the glass is covered than the edge of the glass without covering. The data and results indicate that glass breakage is mainly caused by the temperature difference between the center point and the edge of the glass when heated, and the temperature difference is about 90°C.
This study compares these fire parameters with the types of fire alarm detectors, hoping to find the most suitable detector types for lightweight steel structures, and reduce the fire hazards to personnel.

2. Research Methods

2.1. Fire Dynamics Simulator

The fire simulation software (FDS) used in this study is a calculation developed by the Building and Fire Research Laboratory (BFRL) under the National Institute of Standards and Technology (NIST) The Fire Dynamics Simulator (FDS) developed by Computational Fluid Dynamics (CFD) can simulate a three-dimensional fire scene. Use the Large Eddy Simulation (LES) in the Field Model to simulate the fire scene, divide the indoor space into many small grids, and then use mass, momentum, energy, and combustion The governing equations, such as the conservation of heat transfer, are solved by Navier-Stokes Equation, which can accurately predict the pressure, temperature, velocity, and physical data of smoke flow during a fire. According to Piotr Smardz (2006), "Validation of Fire Dynamics Simuator (FDS) for forced and natural convection flows" master degree, through FDS simulation and small size comparison, it is verified that FDS is correct for forced natural convection flows. Through experiments, it is found that FDS predicts gas flow rate and temperature with an error of 5-20% (related to mass and heat conduction). After calculation, the pressure, Temperature, flow rate, carbon monoxide, carbon dioxide, smoke particles and other important parameters and distribution conditions, using these data can analyze and discuss the situation of the fire scene, suitable for fire simulation of large buildings.

2.2. Initial conditions and boundary conditions

Table 1. Initial conditions and boundary conditions

Item	Initial conditions and boundary conditions
Building size	2000cm(L);600cm(W) ;600cm(H)
Grid Distribution	2000*600*600 (uniform grid point), refer to Bounagui et al. (2003) mentioned that the optimal calculation grid point for house fire is set to 0.2m*0.2m*0.2m in the area outside the fire source.
Initial Temperature	20℃
Space configuration	1F (Front) :factory operation area 1F (Back) :office area 2F : night accommodation
Ventilation state	1F workshop area 1m*1m: 10 locations, 1F office area 1m*1m: 2 locations, Entrance and exit 1m*2m on both sides of 1F: 2 places 2F room 1m*1m:1 place, 2F corridor 1m*1m: 1 place
Wall Material	Exterior wall material: lightweight steel structure Internal wall material: wood
Fire setting	Fire point: below the stairs on the first floor Heat release rate: 2000 kW/m^2 Fire source: gasoline

2.3. Fire detector settings

1. Temperature detection settings (Figure 1):

(1) In the 1F workshop area, a detector, stairs and office are set every 2m.
(2) 2F escape corridor and room.

2. Smoke detector settings (Figure 2):
   Set according to the characteristics of the temperature detector. The setting locations include 1F office; 2F corridor; 2F room to install smoke detectors.

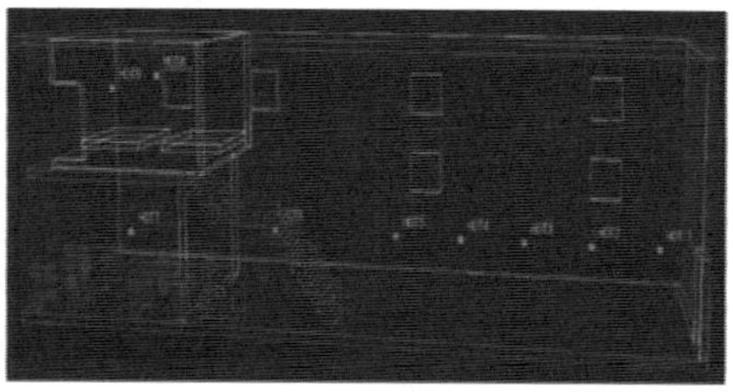

Figure. 1 Temperature detection

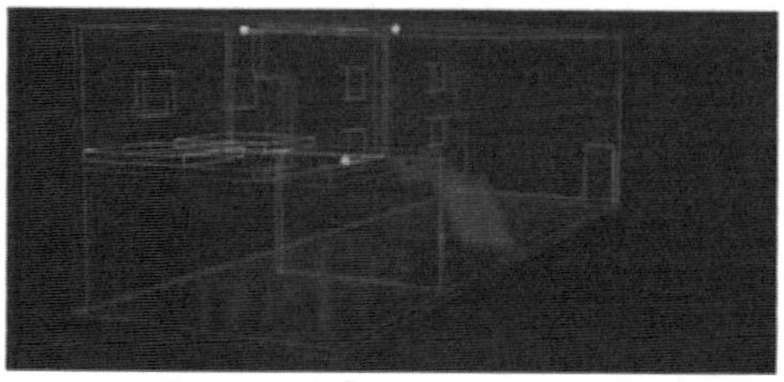

Figure. 2 Smoke detector

3. Results and Discussions

3.1. Temperature Distribution

The thermocouple in the plant area reaches the limit of human body temperature within 8 minutes (Figure 3). The burning time of the walls of the office and factory area is 260 seconds, because the burning is expanded, the temperature of the office and the aisle leading to the office is higher.

After the fire broke out, the 2F corridor reached the highest temperature in about 10 minutes, and the room reached the highest temperature in about 12 minutes. The corridor is close to the fire point, so the smoke accumulation speed and temperature are faster than the room. Therefore, you should escape as soon as possible in case of fire to avoid human body hazards due to high temperature (Figure 4).

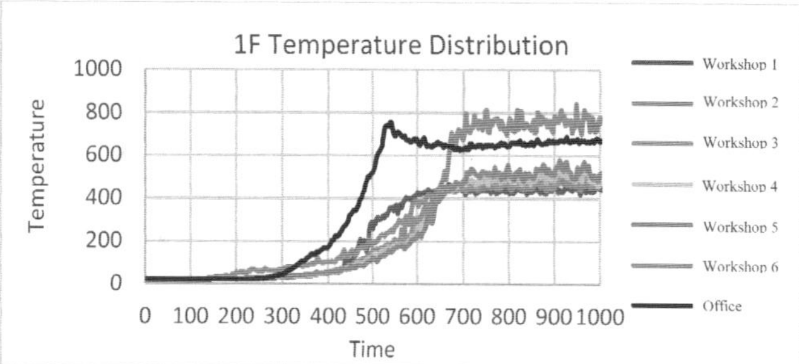

Figure. 3 1F Temperature distribution curve.

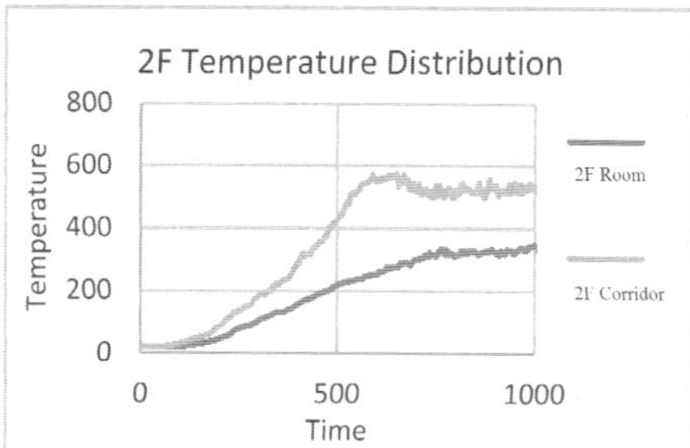

Figure. 4 2F Temperature distribution curve.

3.2. Fire detector alarm time

This study analyzes the starting conditions of the smoke detector and temperature detector. Smoke detection alarm in this model, when the smoke coverage rate reaches 1.68%/m, the alarm device will be activated, because of the thermal buoyancy, the smoke detector in the 2F corridor will alarm at 48 seconds, and the smoke detector in the 2F room will alarm at 80 seconds. The alarm was issued, and the 1F office close to the fire location did not sound the alarm until 118 seconds (Figure5).

A fire alarm will be issued when the temperature detector alarm time is at ambient temperature +30°C. The results of this model show that the temperature detector in the 2F corridor issued an alarm in 185 seconds, and the 2F room had an alarm in 245 seconds. The slowest alarm was issued in the 1F office in 296 seconds(Figure6).

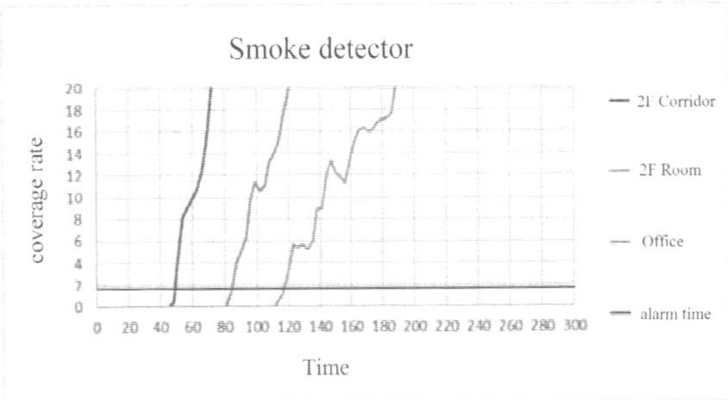

Figure. 5 Fire alarm time- Smoke detector

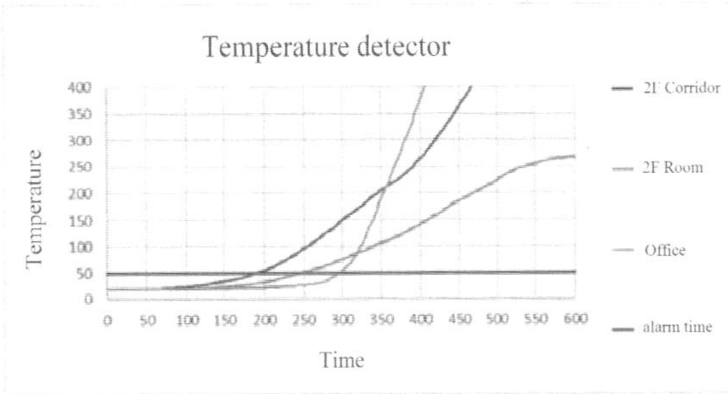

Figure. 6 Fire alarm time- Temperature detector

4. Conclusion and Recommendations

4.1. Conclusion

In this study, FDS was used to simulate the fire of lightweight steel structure buildings. After obtaining the fire environment parameters, the alarm conditions were analyzed with the fire detector. The conclusions are as follows:

1. The fire environment of this building is compared with the living conditions that the human body can withstand as shown in Table 2.
2. In light-weight steel structure buildings, smoke detectors are installed to start alarms 48 seconds after the fire; the fixed-temperature live alarms operate for more than 185 seconds.
3. These factories that store a large amount of combustibles are used as accommodations and have poor evacuation capabilities when a fire breaks out late at night. With the correct fire alarm equipment, the evacuation and initial fire-fighting time can be increased by 137 seconds, and the chance of survival can be increased.

Table 2. Timetable for evacuation

Location	Temperature /Time	Evacuation time
Workshop area	250 /8min	13min
Workspace	250 /7min	12min
Room	250 /10min	15min
Corridor	250 /6min	11min

4.2. Recommendations

1. This research mainly explores the impact of the use of residential fire alarms in lightweight steel structure factories and residential buildings on the lives and property disasters of people. It only simulates two common detectors (temperature and smoke) alarm conditions on the market to estimate the evacuation time.
2. The building model in this study is a hypothetical environment, and the actual scale building can be used to simulate fire and evacuation in the future.

References

[1] Chen Junxun, "Simulation experiment on fire hazard and flash (re)burning of iron houses", research report of the Institute of Architecture, Ministry of the Interior (2001) .

[2] PENG,YI-WEI, "Research on a Detached Buildings "Residential Fire Alarm" Influence of Disaster Reduction" , Master of Civil Engineering Disaster Prevention and Management China University

of Science and Technology, Master Thesis, 2018.
[3] M.J.Skelly, R.J.Roby,and C.L.Beyler,An Experimental Investigation of Glass Breakage in Compartment Fires,J. Fire Protect. Eng.,vol.3,pp.25–34,1991.
[4] Piotr Smardz, 「Validation of Fire Dynamics Simuator (FDS) for Forced and Natural Convection Flows」, University of Ulster Master Thesis,,2006.
[5] Glenn P. Forney,"Smokeview (Version 6) - A Tool for Visualizing Fire Dynamics Simulation Data Volume I: User's Guide, "National Institute of Standards and Technology, NIST Special Publication 1017-1,3-4(2012).
[6] NFAP , Standard Methods of Fire Tests for Evaluating Room Fire Growth Contribution of Textile Coverings on Full Height Panels and Walls, National Fire Protection Association , NFPA265-2007, Quincy, MA, 2007.
[7] Peacock,R.D.,Reneke,P.A.,Bulowaski, R.W.and Babrauskas, V.(1999) Defining flashover for fire hazard calculation. Fire Safety Journal Vol.32.
[8] Spearpoint M, Mowrer FW and Mc Grattan K, "Simulation of a compartment flashover fire using hand calculation, zone model and field model", October 4-8, 1999, Chicago, IL, Society of Fire Protection Engineers, Boston, MA, 3-14 pp, 1999.
[9] Purser D.A.(1995),"Toxicity Assessment of Combustion Products. " SFPE Handbook of Fire Protection Engineering,2nd ed.,pp.2/28–2/146, National Fire Protection Association, Quincy, MA.
[10] Quintiere JG. and McCaffrey BJ, "The burning of wood and plastic cribs in an enclosure", I.NBS (U.S.), NBSIR 80-2054, 1980
[11] Junxun Chen (2001), "Simulation experiment on fire hazard and flash (re)burning of iron houses", Institute of Architecture, Ministry of Interior.
[12] Youqing Zhuang (2004), "Discussion on the Behavior of Steel in High Temperature Environment", Master's thesis of Institute of Civil Engineering, National Cheng Kung University.
[13] Fengya Hong (2009), "Research on the Phenomenon of Iron Sheet Housing in Taiwan", Master's thesis of the Institute of Architecture, National Cheng Kung University.
[14] Hongyi Chen (2016), "Fire Science", Dingmao Bookstore.
[15] Renzhi Chen (2017), "Simulation Study on Fire Rescue Hazards of Residential and Commercial Mixed Tin House Buildings-Taking the Fire Alarm of the Fukano Bakery in Changhua County as an example". Master's thesis of the Department of Environmental Engineering, Daye University.
[16] Qizhong Huang (2018), "Assessing the Relationship between Openings of Lightweight Steel Structure Buildings and Fire Smoke and Heat Hazards" Master's thesis of the Institute of Environmental Engineering, Daye University.
[17] Mingfeng Zhou (2018), "Analysis and Discussion of Fire Alarms for Detached Houses".
[18] Yiwei Peng (2018), "Research on the impact of building "residential fire alarms" in independent buildings on reducing disasters".
[19] Zhongyi Ke (2019) "Fire risk and countermeasures of lightweight steel structure buildings for industrial and residential use".

Numerical analysis of biaxial elasto-plastic buckling of perforated rectangular steel plates applying the Constructal Design method

T da Silveira[1,3], J P S Neufeld[1], L A O Rocha[2], E D dos Santos[3], L A Isoldi[3]

[1] Federal University of Pampa (Unipampa), 810, Tiarajú Av., Alegrete, RS, Brazil.
[2] University of Vale do Rio dos Sinos (Unisinos), 950, Unisinos Av., São Leopoldo, RS, Brazil.
[3] Federal University of Rio Grande (FURG), km8, Itália Av., Rio Grande, RS, Brazil.

E-mail: thiagodasilveira88@gmail.com

Abstract. Thin plates are constantly used in many applications of the structural designs, like naval, aerospace, civil and offshore platforms. When the supported loading is compressive along the plane of plate, it can suffer with an instability phenomenon called buckling which can be elastic or elasto-plastic. In addition, many times the plated structures need to be designed with cutouts for the most varied reasons: access, maintenance, crossing pipes and reducing of weight, for example. The presence of holes affects the mechanical behavior of the plate and it must be investigated. This work used Finite Element Method associated to Constructal Design method and Exhaustive Search technique to find the optimal geometries, i.e. the ones that maximize the Normalized Ultimate Stress, for simply-supported rectangular steel plates under biaxial compressive loading. It was analyzed the size and geometry of the centered elliptical hole. It was observed that the presence of cutout reduces the mechanical strength of the plate and also, the same occurs when the hole size is increased. Constructal Design method has been proved as an effective method to analyze buckling problems conducting to the best geometry for each studied case. For the simulated cases it is evident the relevance of geometric evaluation of elliptical perforations, once the simply variation of geometry can provide differences until around 42% when comparing the optimum and the worst geometries, for the same hole size.

1. Introduction

Thin steel plates are structural components widely used in naval, aerospace, civil and automotive engineering. These elements are characterized by two dimensions on the plane (a and b) much bigger than the out-of-plane dimension, its thickness (t). For specific practical applications these plates must present a cutout insertion (passing pipes, inspection, weight reduction, etc.) modifying its mechanical behavior, due the stress redistribution caused by the presence of perforation [1].

It is well known that an instability phenomenon, called buckling, can occur when plates are under in plane compressive loading. However, plates do not enter in collapse soon after the elastic buckling occurrence, but it can support higher loads without deforming excessively. The load that defines the plate collapse, considering elasto-plastic behavior, is called post-critical or ultimate load (P_u) [2]. When the perforation is centralized to the plate, the buckling load is higher if compared with other hole's positions. Also, the post-buckling analysis shows that for rectangular plates with centralized perforation and biaxial compressive load the buckling resistance is less than a half of the resistance for the uniaxial loading [3]. It is important to mention that elasto-plastic buckling analysis of perforated plates is not a simple task, being the computational modeling via Finite Element Method (FEM) a consistent approach to solve these problems with accuracy.

Given the above, this work seeks to optimize the geometry of a centralized elliptical hole applied to a simply-supported rectangular steel plate subjected to biaxial elasto-plastic buckling. To do so, FEM is used simultaneously with Constructal Design method (CD) and Exhaustive Search technique (ES).

Using degrees of freedom, restrictions and a performance parameter, the CD is applied to define the search space, i.e., to define the different geometric configurations for an engineering system. Based on Constructal Law, this method is the philosophy of evolutionary design in engineering [4, 5]. Thus, all geometries are numerically simulated by FEM and the results compared among each other, allowing the definition of the optimized configuration by means the ES.

2. Buckling of Plates

Considering flat plates, the failure caused by buckling can occurs when it is subjected to compressive loads but, different of columns behavior, for plates it can happens over the critical load of elastic buckling [6-8]. When compressive loads are applied biaxially, the plate buckles with a single half-wave on x and y-direction, being the buckling elastic for slender plates and elasto-plastic for sturdy plates, when the phenomenon occurs on plastic range [9].

The existence of holes in the plate redistribute de membrane stresses, what causes changes in the plate's buckling characteristics and reduce its strength [9]. This redistribution, which is not considered in the theory of small displacements of elastic buckling, usually favors the less rigid regions of the plate and causes an increase in plate efficiency [2].

However, exact and/or experimental analysis of such structural components are complex; especially if a geometrical evaluation is intended to be performed. For this reason, the FEM has been widely used to obtain the ultimate buckling load of plates. To do so, [10] developed design equations to determine the ultimate load for biaxially compressed perforated square plates, through the FEM application. In addition, concluded that the ultimate load capacity is affected significantly by the hole size and the plate slenderness ratio.

It is important to highlight that in the present work the FEM and CD will be applied for the buckling analysis. The use of FEM and CD to evaluate the mechanical behavior of perforated thin plates, subjected to uniaxial elastic and elasto-plastic buckling, was performed by [5] and [11] analyzing different geometries and volume fractions (ϕ) of hole. It was concluded that the redistribution of stresses, caused by the hole, does not affect only the resistance but also the buckling characteristics.

3. Computational Modeling

The computational modeling of the present work used the commercial software ANSYS®, which is based on FEM. The finite element adopted for thin plates analysis is SHELL281, having eight nodes with six degrees of freedom at each node, being: three rotations around x, y and z-axes and three translations in x, y and z-axes [12].

The computational domain for the numerical simulations is depicted in figure 1, showing the reference plate (with no hole) and the perforated plate, both simply supported and submitted to biaxial compressive loading.

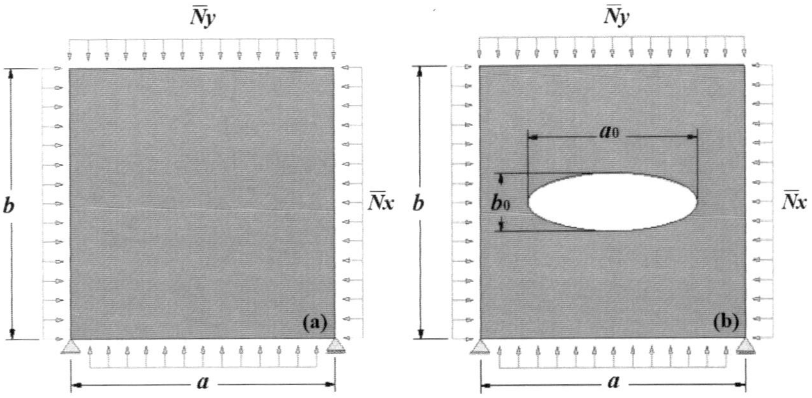

Figure. 1 Simply-supported (a) reference plate and (b) perforated plate, under biaxial compressive loading.

Starting from the first elastic buckling mode configuration an initial imperfection is assumed, being the relation $b/2000$ the maximum value to be assumed. So, the ultimate load of the plate can be found using as reference the load $P_y = \sigma_y.t$, where σ_y represents the material yielding strength [13]. A more detailed explanation about the elastic and elasto-plastic buckling computational modeling can be found in [5].

A computational model validation was made for a simply-supported square plate, with centered circular hole, previously studied by [14]. Regarding figure 1b, a plate with $a = b = 125$ mm, $t = 1.625$ mm, circular hole with diameter $d = 25$ mm and made by steel with $\sigma_y = 323.3$ MPa, the experiment resulted in the ultimate stress $\sigma_u = 73.8$ MPa while the numerical solution resulted in $\sigma_u = 77.6$ MPa. An error of 5.13% was observed comparing results, indicating that the proposed computational model was validated.

4. Constructal Design

The CD is based on Constructal Law, which corresponds to the basis of Constructal Theory and say that "for a finite-size flow system to persist in time (to live), its configuration must evolve in such a way that provides greater and greater access to the currents that flow through it" [15-19]. According to [20], the Constructal Law can be understood as a unifying principle of design.

Regarding the CD application, according to [4], there is no "best" in evolutionary system, but there is "better" today, which turns out to be not as good tomorrow. Design with Constructal Theory is not a mathematical optimization method; then, if the studied system has many degrees of freedom, the CD method can be applied associated with some optimization method, such as the ES technique.

Considering the geometrical evaluation through the CD application, it is necessary to define at least one performance parameter (to be improved), the degrees of freedom (variables), and the geometrical restrictions (fixed parameters). The degrees of freedom are free to vary, since respecting the imposed constraints [5].

In this study the performance parameter is the Normalized Ultimate Stress (NUS), which must be maximized. The NUS factor is obtained through the ratio between the ultimate stress of perforated plate (σ_u) and the ultimate stress (σ_{ur}) of reference plate (without cutout). In its turn, the adopted degree of freedom is the ratio b_0/a_0 (see figure 1b). As restrictions, the dimensions a, b, and t were kept constant in all proposed geometric configurations. In addition, the volume fraction ϕ (defined as the ratio between perforation volume and plate volume) assumed five different values which were kept constant during the cutout's geometries changing.

It is interesting to mention that the CD have been largely employed in Heat Transfer and/or Fluid Mechanics applications, for instance in [21-23]; however one can find only some few works addressed to the application of CD in Structural Engineering, as proposed in the present study.

5. Results and Discussion

First of all, the reference plate (with no hole) was numerically simulated, being a simply-supported rectangular plate with a = 2000 mm, b = 1000 mm and t = 12 mm. The plate submitted to biaxial compressive loading on x and y-direction, is made by AH-36 steel with σ_y = 355 MPa, Young's modulus E = 210 GPa and Poisson's ratio v = 0.30. The ultimate stress obtained through this analysis of σ_{ur} = 596.40 MPa was used to calculate the NUS factor.

In sequence, for the perforated plates, the elliptical hole insertion may respect the volume fractions ϕ = 0.025; 0.05; 0.10; 0.15 and 0.20. For each ϕ, several b_0/a_0 ratios are tested. In order to evaluate the mechanical behavior of the perforated plates, the ultimate stresses σ_u numerically obtained of each proposed geometric configuration was related with σ_{ur}, generating the NUS factor. All geometric configurations were numerically simulated with a converged SHELL281 mesh of 1298 quadrilateral finite elements of 50 mm side, and refined at line around the perforation.

The obtained results for the NUS factor of each studied perforated plate are presented in figure 2.

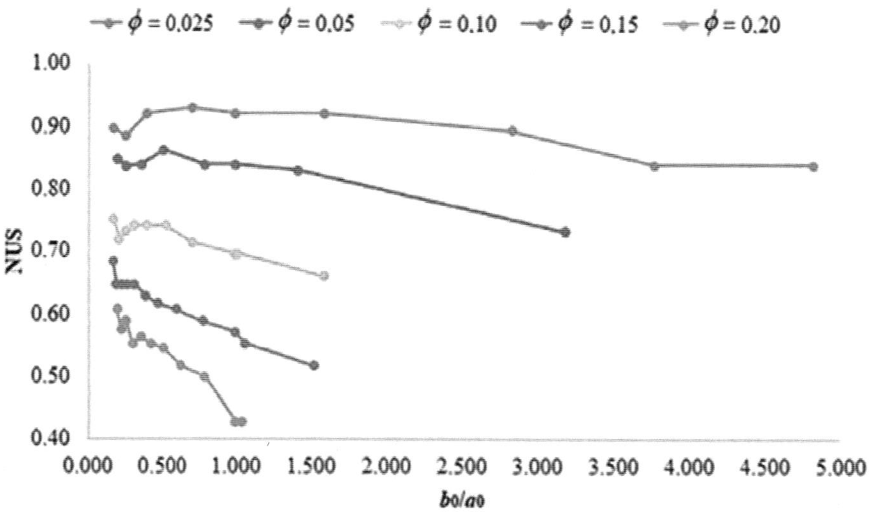

Figure. 2 Obtained NUS for each proposed geometric configuration.

From figure 2 it is possible to observe that, for every simulated ϕ, one maximized NUS is obtained indicating the optimum geometry, i.e., the geometric configuration that provides the higher value to the performance parameter. For ϕ = 0.025 the higher NUS is 0.93, obtained for b_0/a_0 = 0.707. When ϕ = 0.05 the higher NUS achieved is 0.86, for b_0/a_0 = 0.509. In these both cases the optimized geometric configuration is defined by an intermediate value of b_0/a_0. This fact was not repeated for ϕ = 0.10; 0.15 and 0.20, since in these cases the maximized NUS was achieved by geometries with the lower value of b_0/a_0, being, respectively: NUS = 0.75; 0.68 and 0.61; for b_0/a_0 = 0.177, 0.170 and 0.199. From figure 2 one can also infer that as ϕ increases, the magnitude of NUS factor decreases; being an expected trend. Besides, in a general way, the increase of b_0/a_0 ratio promotes a reduction in the NUS value. Finally, for all perforated plates the obtained NUS is smaller than the reference plate, being a coherent behavior as previously noticed by [9].

To understand how the CD defines the optimized geometries, the von Mises stresses distributions are showed in figure 3, for some b_0/a_0 and ϕ = 0.20.

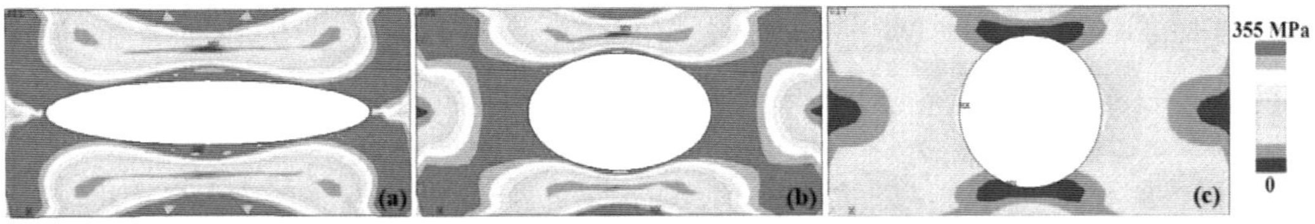

Figure. 3 Distribution of von Mises stresses being (a) the optimum geometry $(b_0/a_0)_o = 0.199$ and maximized $NUS_m = 0.61$ and the geometries with inferior performance (b) $b_0/a_0 = 0.629$; NUS = 0.52 and (c) $b_0/a_0 = 1.039$; NUS = 0.43.

To analyze the von Mises stresses distribution (figure 3), the Principle of Optimal Distribution of Imperfections [19] was applied. In this work, the imperfections are the maximum stresses concentrations and the optimum geometry is the one that provides the best distribution of these imperfections. It is possible to notice that the optimized geometry (figure 3a) has more regions presented in red color, what represents the steel limit stress. In other hand, the other geometries besides having less regions in red color, also, present more areas in green and blue colors that represent the lower stress magnitudes. To the other ϕ, it is worth to highlight that the same trend was noted. Other works did notice analogous behaviors regarding the von Mises stress distribution when analyzing the elasto-plastic buckling of perforated plates [5,24] and stiffened plates [25].

Considering the figure 3, the optimum geometry $(b_0/a_0)_o = 0.199$ (figure 3a) conducted to the maximized performance parameter $NUS_m = 0.61$, while the worst geometry $b_0/a_0 = 1.039$ (figure 3c) conducted to NUS = 0.43. It represents an improvement of 41.86% between the optimum configuration and the worst one for $\phi = 0.20$. Using the same comparison to the other ϕ, the obtained improvements are 10.71%, 17.81%, 13.64% and 30.77% for ϕ = 0.025; 0.05; 0.10 and 0.15, respectively.

6. Conclusion

The obtained results showed that the perforation's geometric evaluation is an important analysis to be done once it can vary the plate's mechanical behavior under elasto-plastic buckling. Through the Principle of Optimal Distribution of Imperfections, the CD method, associated to ES technique and FEM, proved to be an effective tool for biaxial elasto-plastic buckling analysis once it has conducted to the optimum geometry which provided the highest performance parameter.

Furthermore, it is important to mention that was possible to obtain the limit stresses for each perforation geometry and volume fraction, from which might cause the collapse of plate submitted to biaxial elasto-plastic buckling. For all simulated cases the relevance of geometric evaluation of the perforations is evident, since the simply variation of geometric configuration can provide improvements of 10.71%, 17.81%, 13.64%, 30.77% and 41.86% for ϕ = 0.025; 0.05; 0.10; 0.15 and 0.20, respectively.

Acknowledgements

The authors thank to Federal University of Rio Grande - FURG and to Federal University of Pampa (Unipampa). Moreover, the authors L. A. O. Rocha, E. D. dos Santos and L. A. Isoldi thank CNPq for their research grants (Processes: 307791/2019-0, 306024/2017-9, and 306012/2017-0, respectively).

References

[1] Shojaee T, Mohammadi B, Madoliat R and Salimi-Majd D 2019 Development of a finite strip method for efficient prediction of buckling and post-buckling in composite laminates containing a cutout with/without stiffener *Composite Structures* **210** pp 538–552
[2] Chajes A 1974 *Principles of Structural Theory* (New Jersey: Prentice-Hall Englewood Cliffs)
[3] Mohammadzadeh B, Choi E and Kim W J 2018 Comprehensive investigation of buckling behavior

of plates considering effects of holes *Structural Engineering and Mechanics* **68(2)** pp 261–275
[4] Rocha L A O, Lorente S and Bejan A 2017 Constructal Theory in Heat Transfer. *Handbook of Thermal Science and Engineering* ed 1 **1** pp 1–32.
[5] Helbig D, Silva C C C D, Real M D V, Dos Santos E D, Isoldi L A and Rocha L A O 2016 Study About Buckling Phenomenon in Perforated Thin Steel Plates Employing Computational Modeling and Constructal Design Method *Latin American Journal of Solids and Structures* **13(10)** pp 1912–1936.
[6] Jones R M 2006 *Buckling of bars, plates and shells* (Blacksburg: Bull Ridge Publishing)
[7] Szilard R 2004 *Theories and Applications of Plate Analysis: Classical Numerical and Engineering Methods* (New Jersey: John Wiley & Sons)
[8] Trahair N S and Bradford M A 1988 *The behavior and design of steel structures* (New York: Springer)
[9] El-Sawy K M and Martini M I 2010 Stability of biaxially loaded square plates with single central holes *Ship and Offshore Structures* **5(4)** pp 283–293.
[10] Shanmugam N E, Thevndran V and Tan Y H 1999 Design formula for axially compressed perforated plates *Thin-Walled Structures* **34** pp 1–20
[11] Helbig D, Cunha M L, Da Silva C C C, Dos Santos E D, Iturrioz I, Real M De V, Isoldi L A and Rocha L A O 2018 Numerical study of the elasto-plastic buckling in perforated thin steel plates using the constructal design method *Research on Engineering and Structural Materials* **4(3)** pp 169–187.
[12] Ansys Mechanical APDL 2015 *Version 15.0 – User's Guide* (USA: ANSYS Inc)
[13] El-Sawy K M, Nazmy A S and Martini M I 2004 Elasto-plastic buckling of perforated plates under uniaxial compression *Thin-Walled Structures* **42(8)** pp 1083–1101
[14] Narayanan R and Chow F 1984 Strength of Biaxially Compressed Perforated Plates *Int. Specialty Conf. on Cold-Formed Steel Structures* pp 55–73
[15] Bejan A 2000 *Shape and Structure, from Engineering to Nature* (Cambridge: Cambridge University Press)
[16] Bejan A and Lorente S 2004 The constructal laws and the thermodynamics of flow systems with configuration. *Int. J. Heat Mass Transfer* **47** pp 3073–3083.
[17] Bejan A and Lorente S 2005 *La Loi Constructale* (Paris: L'Harmattan)
[18] Bejan A and Lorente S 2008 *Design with Constructal Theory* (New Jersey: Hoboken).
[19] Bejan A and Zane J P 2012 *Design in nature: how the constructal law governs evolution in biology, physics, technology and social organization* (New York: Doubleday)
[20] Rocha L A O, Lorente S and Bejan A 2013 *Constructal Law and the unifying principle of design* (New York: Springer)
[21] Hazarika S A, Bhanja D and Nath S, 2020 Fork-shaped Constructal fin array Design a better alternative for heat and mass transfer augmentation under dry, partially wet and fully wet conditions *International Journal of Thermal Sciences* **152**, pp 106329.
[22] Ganjehkaviri A and Mohd Jaafar M N 2019 Multi-objective particle swarm optimization of flat plate solar collector using Constructal Theory *Energy* **194** pp 116846.
[23] Hermany L, Lorenzini G, Klein R J, Zinani F F, Dos Santos E D, Isoldi L A and Rocha L A O 2018 Constructal Design applied to elliptic tubes in convective heat transfer cross-flow of viscoplastic fluids *International Journal of Heat and Mass Transfer* **116** pp 1054–1063.
[24] Lorenzini G, Helbig D, Da Silva C C C, Real M V, Dos Santos E D, Isoldi L A and Rocha L A O 2016 Numerical evaluation of the effect of type and shape of perforations on the buckling of thin steel plates by means of the Constructal Design method *International Journal of Heat and Technology* **34** pp S29–S20.
[25] Lima J P, Cunha M L, Dos Santos E D, Rocha L A O, Real M V and Isoldi L A 2020 Constructal Design for the ultimate buckling stress improvement of stiffened plates submit-ted to uniaxial compressive load *Engineering Structures* **203**.

A novel, structurally coloured coating for aluminium alloys

Ian J Mardon[1], Fengyan Hou[1], and Chris Goode[1, 2]

[1] Cirrus Materials Science Ltd, Auckland, New Zealand
[2] Author to whom any correspondence should be addressed.

Abstract: Current approaches to coat aluminium surfaces that use paints and organic coatings to protect them from corrosion and provide an aesthetic finish, are also energy intensive and involve high emissions of VOCs. The use of these coatings is becoming increasingly restricted by global efforts towards more environmentally sustainable surface finishing technologies. A recent development based on porous anodised aluminium (PAA) and metal nanowire deposits is presented here, which may become a competitive alternative to paint and organic coatings. Anodising and electrodeposition do not emit VOCs and are less energy intensive. By controlling the conditions for anodising and metal electrodeposition, coatings in a wide spectrum of colours can be produced, including greys, black, blues, purples, and greens. Characterisation of the coating morphologies and corresponding colours agrees with simulated results based on the Fresnel equations and Bruggeman Mixed Media Approximations.

1. Introduction

Aluminium alloys are a common choice of material in applications, such as in architecture, automotive, and aerospace, due to their desirable strength-to-weight ratio and their corrosion resistance oxide layers. Previously, a variety of anodised surfaces with metal deposition has been used to create coloured surfaces on light metal substrates [1], mainly the electrolytic colouring of Aluminium alloys. These approaches, however, have typically used sulphuric acid anodising with AC electrolysis with metal salts, resulting in colour generation that is purely dependent on light interference behaviour within the anodising pores.

Cirrus Hybrid™ is a patented technology that combines anodised light metal surfaces with functional metal and/or polymer materials. In this paper, a Cirrus Hybrid™ coating is presented, whereby nickel nanowires are deposited into PAA. This combination results in a coloured coating, where the optical properties are due to the coating morphology. Here we report both a novel process and a mathematical model for the coating system with the ability to produce a full spectrum of coloured surfaces. Modelling of the coating was performed through use of the Fresnel equations and Bruggeman Mixed Media Approximation, coupled with measurements of the coating morphology. Simulations based on the theory and measurements of the coatings show good agreement with colour measurements of the coatings. Furthermore, these results suggest that the critical colour determinant is the barrier layer thickness and PAA allows very thick barrier layers to be developed [2]. To support thick barrier layers without the associated pore widening, which would weaken the coating and produce an easily corroded surface, the anodizing bath must be carefully controlled [3][4]. Modelling and experimentation also suggest that the metal/air interface above the deposited nanowires is vital for retaining the coating colour; thus, in order to produce a surface that is industrially viable, specific seal characteristics are required.

Here we also report on two interesting technologies to produce such a seal. With the correct choice of sealing material, this coating technology has great potential. In many current applications that use

IOP Conf. Series: Materials Science and Engineering 1048 (2021) 012018 doi:10.1088/1757-899X/1048/1/012018

organic-based colour coatings, which produce environmentally damaging and toxic wastes and by-products, such as Volatile Organic Compounds (VOCs), this Cirrus Hybrid™ colour coating may be an effective replacement. The authors posit that such replacement would see decreased carbon emissions, decreased VOC emissions, as well as cost savings from the low-energy application, and greater ease of recycling for end-of-life processing.

2. Materials and Methods

2.1 Materials

Aluminium Al-6061 alloy sheet, 2mm thick, was laser cut into 3cm x 5cm coupons. A 2.5mm hole was drilled through the top of each sample to accept 2.5mm aluminium wire via friction fit to provide electrical connection to the coupons, and to hold them in place during treatments. All solutions were prepared with deionised water. 50% v/v Nitric Acid, Probright, and Metaclean ZX-E were purchased from PPS Industries (NZ) for pre-treatment of the coupons. A proprietary, phosphoric acid-based anodising bath was used for the anodising of the aluminium coupons. Polyethylene glycol was used as an anodizing bath additive to allow for higher voltage anodizing. AR grade nickel chloride and boric acid, and LR grade nickel sulphate were purchased from ECP (NZ) for the formulation of a Watts-type commercial nickel electroplating bath.

2.2 Methods

Cross-sections were metallographically prepared and viewed under SEM (either a Philips CL30S FEG or a Hitaches SU-70 Schottky Field Emission) and optical microscope (Phenix, with Phmias image capture software). Cross sections allowed for measurement of coating thicknesses as well as morphology of the coatings. The substrates' surface roughness (pre-anodising) was obtained using a Bruker Optical Surface Profiler.

The colour of the colour coatings were measured by analysis in Imagej software, of images taken by a digital camera of the coupons in a light box. Imagej was also used to take measurements from microscope images. Spyder 4.1.3 (with NumPy, SciPy, and mpmath packages) was used to generate coating colours from simulated reflectance spectra which were calculated from the coating morphology and optical properties using Fresnel's equations and Bruggeman's mixed media approximation, as discussed further in section 3.5.

To test the performance of the coatings, ASTM B571 was followed to test the coating adhesion, and ASTM B117 was followed, using a 5% wt. sodium chloride salt solution to evaluate the corrosion resistance.

3. Experimental

3.1 Substrate Preparation

The aluminium coupons were polished, degreased in Metaclean ZX-E solution, followed by chemical activation of the surface using Probright acid etch. A 50% Nitric Acid dip for several minutes was used to de-smut the substrates prior to anodizing.

3.2 Application of Cirrus Hybrid™ colour coatings

Duplicate samples were prepared for each of the chosen parameters to allow for independent testing of the coatings' properties. The coupons were anodised in a proprietary anodising bath at 60, 70, 80, and 90 volts, to evaluate the effect of anodising voltage on the resultant colour coating. The anodising temperature was maintained at room temperature by using a recirculating cooler, with air agitation of the anodising solution.

Anodised samples were rinsed in deionised water and submerged into the electroplating bath for a period to allow the solution to diffuse into the porous surface. An initially low current density was applied to begin electrodeposition, allowing for nucleation of electrodeposited metal at the base of the pores. The current density was increased to allow for the formation of nanowires, only partially filling the porous surface. The colour coatings were then rinsed with deionised water, and dried using compressed air. Direct and Pulse Current (DC and PC) electrodeposition were compared to evaluate the effect on the resulting colour uniformity.

3.3 Sealing
It was observed that the coatings after electrodeposition lacked corrosion resistance. The close contact between the electrodeposited nickel and the aluminium allows for galvanic corrosion. UV-initiated and electrolytically prepared polymer seals were applied to improve corrosion resistance.

3.4 Testing coating performance
Sealed coatings, with unsealed control samples, were subjected to neutral salt spray corrosion testing, as per ASTM B117. The samples were observed every 12 hours, and the corrosion points were counted. The number of corrosion points over time was used as a metric to analyse the corrosion performance of sealed and unsealed coatings.

Duplicate samples were subjected to the mechanical testing methods to evaluate the performance of the coatings. The measurements from these mechanical tests were recorded, and compiled to evaluate the potential performance of these coatings for commercial applications.

3.5 Evaluating coating morphology and colour
Duplicate samples were imaged using SEM and optical microscopy. The resulting images were used to take measurements of the pore diameter and interpore distance of the main and side pores observed in the anodised layer. The coating thicknesses were also assessed. The coating was simulated as a stack of layers of alternating porosity, where the high porosity layers were the diameter of the side pores, and the low porosity layers were the inter-pore spacing of the side pores. The calculated refractive index for the Alumina-nickel and Alumina-Air Bruggeman mixed media then allowed for determination, using the Fresnel equations, of simulated reflectance spectra. These spectra were integrated with CIE colour matching functions to produce an RGB value associated with a coating morphology. These colour values were compared to colour measurements of the samples.

Samples were photographed in a lightbox, and the images processed in Imagej to evaluate the colour (in the CIE-Lab and -RGB colour coordinate systems) of each sample, as well as the internal colour variance, measured by ΔE

$$\Delta E = \sqrt{(\sigma_R)^2 + (\sigma_G)^2 + (\sigma_B)^2} \quad (1)$$

Colour difference between samples and their simulated colours were evaluated using equation (1), except the difference between Lab colour coordinates was used in place of standard deviations.

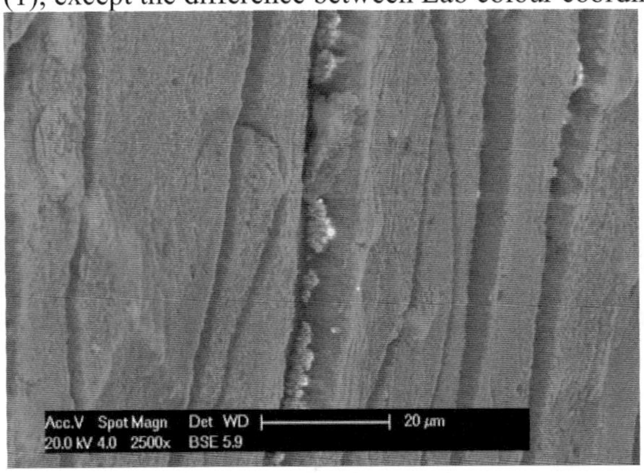

Figure. 1 The open, anodised aluminium oxide is dull compared to the high conductivity of the nickel, shown as bright deposits in the oxide

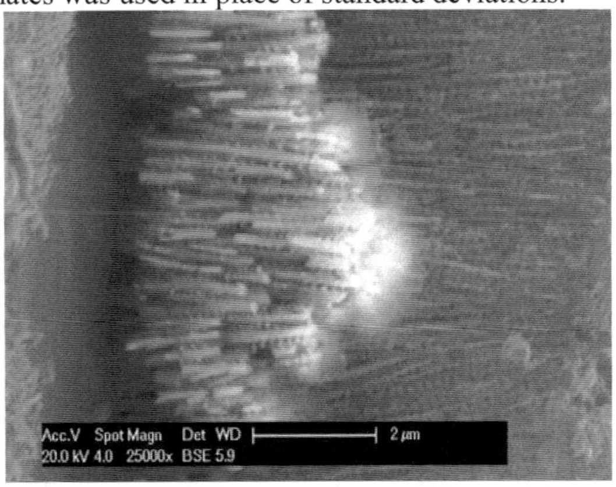

Figure. 2 High magnification of the centre of Figure 1.

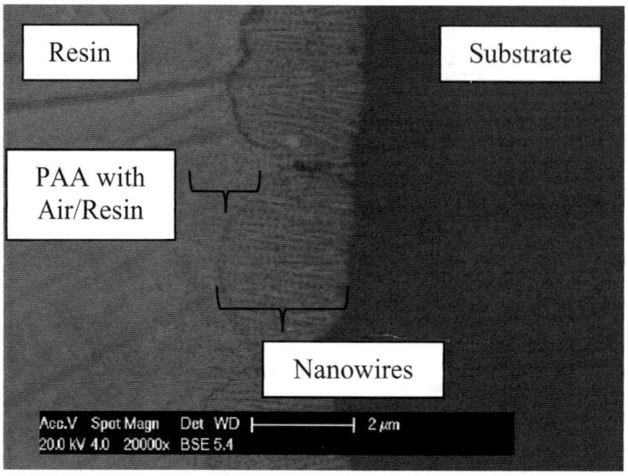

Figure. 3 SEM image of a metallographic cross section, with (left-to-right) resin, unfilled oxide, nickel filled oxide, and substrate.

4. Results and Discussion

4.1 SEM and Optical Microscopy

Figure 1 shows an SEM of a hybrid film that has been bent to expose the metal nano rods. Here the regular arrangement of vertical pores may be clearly observed. Figure 2 depicts the metal nanorods in the coating. This deposit was created using direct current, and uneven pore filling [5] resulting from uneven nucleation across the pores of the surface. The differing nanowire lengths produce a spread in the resulting colour spectrum and produce muddy colours. Pulse deposition develops more uniform nanowire length and brighter colours. The side pores which periodically connect the main pores with the oxide layer may be observed in Figure 2. These side pores are approximately 150nm from centre-to-centre, and are 70nm in diameter, at their opening to the main pores. Figure 3 further shows the pore filling where the interlinked nanowires are of varying lengths due to the use of DC electrodeposition. This characteristic porosity was shown to be an important component in producing a specific colour, when attempting to model the optical properties of the coating.

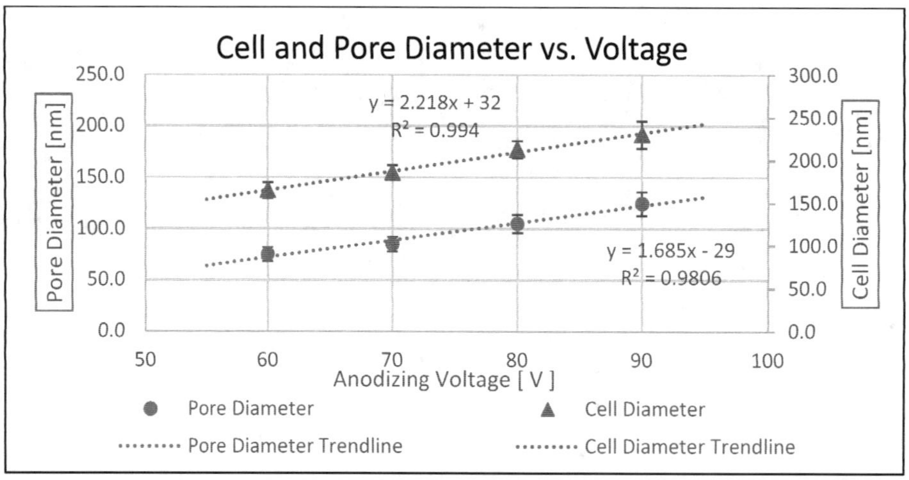

Figure. 4 Trends of Pore and Cell diameters vs. Anodizing Voltage.

Figure 4 shows the relationship between Pore and Cell average diameter and the anodizing voltage. The linear trend is typical of acid bath anodizing. It is important to note that the cell diameter increases faster than the pore diameter, meaning the porosity of the coating decreases with increasing Voltage. For mechanical properties related to porosity, this would imply an increase in toughness with Voltage, however this is outside the scope of this paper. The decrease in porosity also signals a lower volume fraction of nickel in the alumina matrix; nickel is a very lossy material in the visible spectrum, so less nickel will mean less absorbance, as compared to the dielectric alumina.

4.2 Initial Modelling of Sample Morphology

Through use of Bruggeman Mixed Media approximation [6] and the Fresnel Equations [5], the reflection spectra of simplified coatings were simulated. The detailed results, which will be available in a future publication, show good agreement between the simulation results and the measurements of prepared samples. These results demonstrate that the colour generation technology is a unique result of application of Cirrus Hybrid™ with a colouring functional layer.

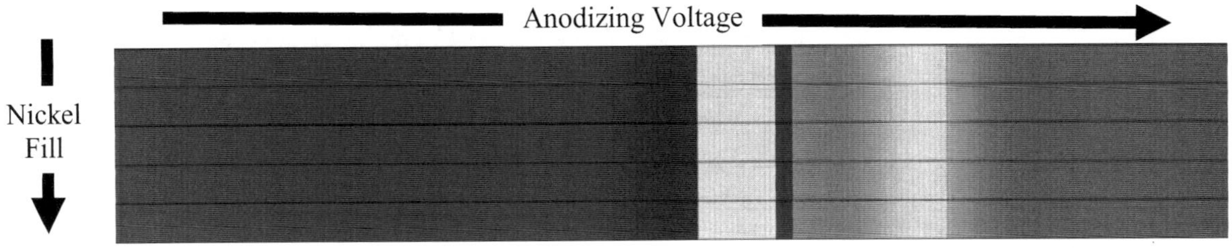

Figure. 5 Simulation results from measurements of coating morphology, Bruggeman mixed media approximation, and the Fresnel equations. NB: reader to refer to online version for colour.

Figure 5 shows the results of the optical modelling of the coating morphology and composition. The CIE colour matching functions are designed to produce colour coordinates for reflection spectra. In the left of Figure 5, the brown colour correlates to black on real samples, which implies that the resonant peak in the reflection spectra is outside the visible spectrum. The colours observed, left-to-right, show that black moves towards dark blue, lightening in colour with a section of light blue, before moving into yellows, greens and reds.

4.3 Colour Analysis

The prepared coatings were imaged, and the colour coordinates and colour variation (ΔE) were evaluated, as described in Section 3.5. CIE defines a ΔE of 2.3 to be the "Just Noticeable Difference" [7], and many paints are designed to produce coatings with ΔE less than 1.

Table 1. RGB values and colours, as well as internal colour variation for samples at 60V, 70V, and 90V.

Anodizing Voltage (V)	Colour	R	G	B	RGB	Lab ΔE
60	Light Grey	130	132	139		1.9
60	Dark Grey	73	75	81		1.3
60	Black	37	44	52		1.1
70	Dark Blue/Purple	48	61	92		1.4
80	Dark Blue	44	72	125		1.6
90	Blue	44	76	144		1.5

The observed colour varied little as the nickel electrodeposition time (and thickness) increased for samples anodised at the same voltage. Samples anodised at the same temperature were seen to shift from black to dark blue, to light blue, as the anodizing voltage increased as shown in Table 1. Table 1 also shows three samples anodised at 60V, with increasing anodizing temperatures. The increased temperature allows for more electrolyte dissolution of the porous anodised oxide, resulting in higher porosity, and thus darker colours due to more light absorbing nickel in the coating.

There is good agreement between the simulation results (4.2) and the real world colours, thus we may conclude that the thickness of the barrier layer, with minor filtering properties of the air- alumina-mixed layers above, gives rise to the coating colour.

4.4 Corrosion Performance of Coating

Samples were subjected to Corrosion Resistance testing, as per ASTM B117. Unsealed coatings typically survived no longer than 48 hours, before corrosion points occurred, which required the development of a sealing layer to protect the underlying colour coating.

The inherent galvanic corrosion between the nickel electrodeposits and the Aluminium required the investigation of appropriate sealing material, to avoid the contact between the nickel deposits and corrosive media.

Table. 2 Corrosion resistance as well as the colour change between unsealed and sealed samples.

Sample	Hours until corrosion	Seal colour change
Unsealed Control	48	n/a
UV Polymer Seal	336	5.1
Electropolymer Seal	336	10.7

Table 2 shows the effect of polymer seals on the corrosion resistance, where both polymer seals provided approximately an 8-fold improvement. Both seals also significantly changed the perceived colour of the coating. The electropolymerized seal layer, whilst only ~0.2 μm thick, provided a uniform barrier layer between the corrosive media, the underlying nanorod and substrate. This seal, however, was not optically clear, so the colour of the coating was masked by its dark brown/black colour. The UV polymer also provided an effective barrier layer, with corrosion points only observed at the edges of the samples where the monomer had not been applied. Whilst the UV polymer preserved the coating colour better, there was still a change in colour. Modelling of the colour changes due to these polymer seals showed that a change in refractive index of the unfilled pores of the coating saw a colour change similar to that observed. This result suggests that air or vacuum is needed in the pores to preserve the colour.

5. Conclusion

The development of this novel colour coating has allowed for the integration of electrodeposition and proprietary anodizing to produce coatings in black, shades of grey, and various shades of blue on aluminium alloys. Making use of this robust, alternative coating technology, in combination with an appropriate sealing system, may offer a competitive replacement for paint. The ability to tune the anodised layer morphology allows for control over the resultant colour. The Cirrus Hybrid™ colour coating system is therefore a lightweight, low-energy application, which does not require the use of VOCs or other toxic, environmentally unfriendly compounds.

6. References

[1] Sato K T, and Yasuhiko I 1981 Observation of barrier layer during ac electrolytic colouring of anodized aluminium *Electrochem. Acta* **26 (9)** 1-4

[2] Eftekhari A 2008, *Nanostructured Materials in Electrochemistry*, John Wiley & Sons, pp 14-18

[3] Faucheu J, Matsapey N, Charriere R, Flury M and Delafosse D 2015 Structural Colors of Nanoporous Anodic Alumina: Overview of Recent Advances and Case Study in Elaboration, Characterization, Photometry and Modelling *Current Nanoscience* **11** pp 317-325
[4] Yisen L, Yi C, Zhiyuan L, Xing H 2011 Structural coloring of aluminum *Electrochem. Comms.* **11** pp 1336-1339
[5] Ruiz-Clavijo A, Tsurimaki Y, Caballero-Calero O, Ni G, Chen G, Boriskina S V and Martin-Gonzalez M 2018 Engineering a full gamut of structural colors in all-dielectric mesoporous network metamaterials *ACS Photonics* **5** pp 2120-2128
[6] Macleod H A 2017, *Thin-film optical filters: series in optics and optoelectronics*, CRC Press Ltd, pp 13-103
[7] CIE Technical Committee 1-48, *CIE Technical Report: Colorimetry 3rd* Edition, 2004 International Commission on Illumination (CIE)Faucheu J, Matsapey N, Charriere R, Flury M and Delafosse D 2015, *Current Nanoscience* **11** pp 317-325

Effects of the Carbon Fiber Orientations for the Strengthening of Thin-Walled Steel Cylinders under Compressive Loads and Bending Shear Loads

P V Nhut[1], Y Matsumoto[2], T Matsui[3], and H Nakamura[4]

[1] Graduate student, Department of Architecture and Civil Engineering, Toyohashi University of Technology, Tempaku-cho, Toyohashi, 441-8580 Japan
[2]Associate Professor, Department of Architecture and Civil Engineering, Toyohashi University of Technology, Tempaku-cho, Toyohashi, 441-8580 Japan
[3]ACM Technology Dept., TORAY Industries, Inc., Nihonbashi-muromachi, Chuoh-ku, Tokyo, 103-8666, Japan
[4]Associate Professor, Department of Civil and Environmental Engineering, Tokyo Metropolitan University, Minami-Osawa, Hachioji-shi, 192-0397 Japan

Email: phan.viet.nhut.yu@tut.jp

Abstract. In this study, the effects of carbon fiber orientations on the strengthening of small-scale thin-walled steel cylinders (TSCs) under axial compression were experimentally and numerically investigated. Besides, the strengthening effects of various types of carbon fiber reinforced polymers (CFRPs) on the load-carrying capacity of full-scale TSCs were numerically investigated. The results showed that circumferential CFRP layers (0°) have significant effects on the increased load-carrying capacity of TSCs under axial compression, whereas, only 90° CFRP layers had small strengthening effects for TSCs under axial compression. When the TSCs subjected to bending shear loads with high values of internal pressures, the appearances of circumferential fibers also brought better strengthening effects than only the angled fibers for TSCs.

1. Introduction

Large thin-walled steel cylinders (TSCs) are used as fluid tanks, chemical, electric power, food engineering, etc., in many fields of industry and infrastructures. Many of them were constructed a long time ago and suffered degradation because of aging or corrosion. Moreover, a large number of earthquakes with highly recorded magnitudes in recent years, especially in the Pacific Ocean, threaten the safety of not only degraded tanks but also existing healthy tanks [1, 2]. As a result, the strengthening of the tanks against buckling and enhance the ultimate strength of the tanks is necessary to improve performance against seismic events. Recently, carbon fiber reinforced polymers (CFRPs) that are lightweight, very strong, and are resistant to corrosion have been used to increase the load capacity and ductility of steel structures, thus proving to be an economical and reliable strengthening solution [3-7]. However, the study about the use of CFRP for the strengthening of TSCs is limited. Therefore, clear research about the strengthening effects of CFRP on the load-carrying capacity of TSCs is the important work to develop this application.

Some research has been conducted to investigate the strengthening effects of bonded CFRP layers on the increase of the load-carrying capacity of TSCs [8-11]. All the above research focused on the use of circumferential CFRP layers for the strengthening of TSCs under axial compression. The results showed that circumferential CFRP layers can restrain the elephant foot bucking (EFB) at the base of TSCs and increase the load-carrying capacity of TSCs. Nhut et al. [12] experimentally investigated the effects of circumferential CFRP layer (0°) and bi-directional CFRP layer (0°/90°) on the strengthening of TSCs against dynamic loads. They concluded that the 0° CFRP layer has better effects than the 0°/90° CFRP layer under dynamic loads. In this study, the effects of carbon fiber orientations on the strengthening of small-scale TSCs under axial compression were experimentally and numerically investigated. Moreover, finite element analysis (FEA) was performed to analyze the strengthening effects of different types of CFRP layers with various fiber angles on the load-carrying capacity of full-scale TSCs under bending shear loads. The use of FEA had been proven to bring good correspondence between experimental and numerical results [11, 13]. Therefore, FEA is a suitable solution to investigate the strengthening effects of CFRP layers for full-scaleTSCs under bending shear loads.

2. Carbon fiber sheets and resin

Two kinds of carbon fiber (CF) sheets, products of Toray Industries, Inc., were employed: unidirectional CF (UT70-20) and bi-directional CF (BT70-20). The CF UT70-20 was used to strengthen in 0° (UT0) or 90° (UT90), whereas, CF BT70-20 was used for 0°/90° strengthening (BT0/90) or ±45° strengthening (BT45). ±45° CF sheets were applied by rotating the 0°/90°CF sheets to 45°. The CFRP-strengthened specimens were molded following the next two steps: First, we applied primer resin (E810LS, Konishi, Osaka, Japan) for the strengthened positions and waited around 24 hours for the curing. Second, the CFRP layers were bonded and molded on the outside surfaces of TSCs using impregnation epoxy (E2500-Konishi, Osaka, Japan) by hand lay-up molding method. E2500 has elastic modulus and Poisson's ratio values of 4.1 *GPa* and 0.37, respectively.

The CF sheets were bonded onto the outside surfaces of TSCs for strengthening. The CF sheets after impregnating with E2500 resin created CFRP layers. The material properties of one CFRP layer were calculated from the lamination theory, as shown in Table 1. In the table, t_f and t_c are the thickness of carbon fiber and CFRP layer, respectively; V_f is the volume fraction of carbon fiber and epoxy; E_x and E_y are the elastic modulus of CFRP layers in x and y direction (circumferential (0°) and vertical direction(90°)); G_{xy} is the shear modulus of CFRP layer.

Table 1. The thickness and mechanical properties of one CFRP layer.

Type	CFRP layer	Carbon Fiber Orientation	t_f (mm)	t_c (mm)	V_f (%)	E_x (MPa)	E_y (MPa)	G_{xy} (MPa)	Poisson's ratio
A	UT0	0°	0.222	0.63	35	88415	6251	2283	0.3
B	BT45	±45°	0.224	0.64	35	8360	8360	22874	0.83
C	BT0/90	0°/90°	0.224	0.64	35	47562	47562	2283	0.04
D	UT90	90°	0.222	0.63	35	6251	88415	2283	0.3

3. Investigation models

Figure 1 shows two types of models using for the investigation. Small-scale TSCs were subjected to compressive loads, whereas, full-scale TSCs were subjected to bending shear loads (horizontal shear load and inside internal pressures). In the figure, h is the height of TSC, h_{CF} is the height of the strengthened CFRP layer, d is the outside diameter of TSC, and l is the distance from the base of TSC to the point that horizontal loading is applied (top of the rigid roof).

The effects of carbon fiber orientations on the strengthening of small-scale TSCs were investigated by both experiments and FEA. Three different types of CFRP layers, including *Type A, C*, and *D*, were used to consider the effects of carbon fiber angle for the small-scale models. The small-scale TSCs used in the experiments were machined from the seamless mild steel pipes (STKM13A, the original outside diameter of 267.4 mm and original thickness of 8 mm). After machined by numerically controlled lathe, these TSCs had 200 mm heights, 2.0 mm thicknesses, and 260mm outside diameters. The material of small-scale TSCs had an average yield tress of 286 (MPa), which was determined from material tests. Two CFRP layers were covered within 140mm height at the bottom of the TSCs; while three CFRP layers were covered within 50mm height at the top of the TSCs. The purpose of this work was to prevent the elephant foot bulge (EFB) occurrence at the top of the TSCs. EFB was a main failure mode of TSC under compressive loading. In the experiment, the EFB could occur at the top or bottom of TSCs if the same CFRP layers were applied on the TSC's walls because symmetrical loading conditions were applied at the top and bottom of TSCs. Therefore, the covering of three CFRP layers could prevent the EFB occurrence at the top of TSCs, making the investigation of the effects of carbon fiber orientations easier.

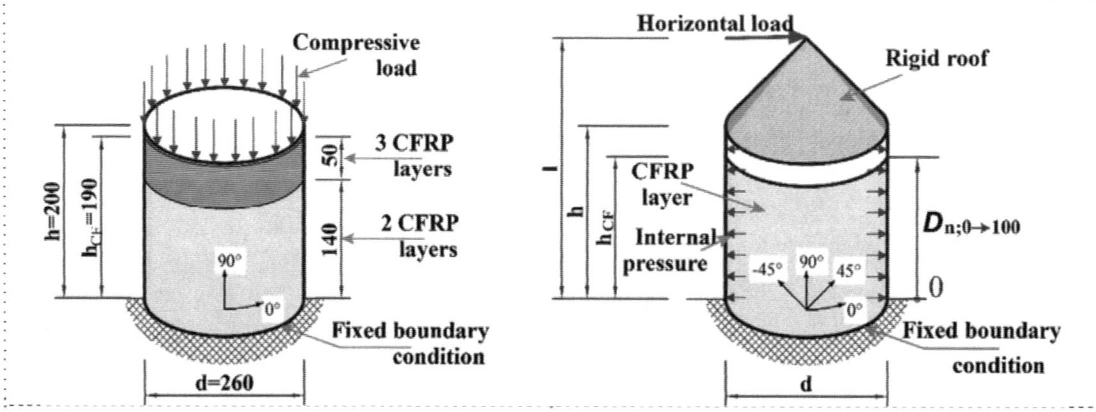

(a) Small-scale models subjected to a compressive load (b) Full-scale models subjected to bending shear load

Figure. 1 Study models. Unit: mm.

For the second type model, full-scale TSC with the dimension *l/h/r/t* (mm) = 6200/5000/4990/10 was used for the strengthening of CFRP layers. Here, *r* is the inside radius of TSC and *t* is the thickness of TSC. The materials of the full-scale TSCs are assumed using SS400 steel (Japanese Industrial Standard) with a yield stress of 235 MPa and an elastic modulus of 205 GPa. FEA was used for the analysis of 36 models with different types of CFRP layer strengthening, which were divided following the notation $D_{n;0\rightarrow100}$. Figure 2 describes the parameters of this notation. Internal pressures were applied onto the inside surfaces of TSCs. Two types of internal pressures were assigned in the analysis with the ratios $\sigma_h/\sigma_y = 0.5$ and $\sigma_h/\sigma_y = 0.7$. Here, σ_h is the tensile hoop stress and σ_y is the yield stress of SS400 steel. From the tensile hoop stress σ_h, the values of the internal pressures are defined by equation (1) with r_1 is the mean radius of the TSC. Three types of CFRP layers: *Type A*, *B*, and *C* with different orientations of carbon fibers were used to strengthen the TSCs. Every type of CFRP layers was used to strengthen 20%, 50%, and 100% of the height of TSC and subjected to two levels of internal pressures. Tables 2 and 3 show the kinds of TSCs analyzed by FEA.

$$p = \sigma_h t / r_1 \tag{1}$$

$D_{n;\,0\rightarrow100}$ ← Height percentage of strengthened CFRP layers from the bottom (0) to full height of TSC (100)
The numbers of CFRP layers
The direction of the CFRP layers (D= 0°; ±45°; 0°/90° or D=Type A; Type B; Type C)

Figure. 2 Explanation of the types of strengthening CFRP layers.

4. Effects of the angles of carbon fibers on the strengthening of TSCs under compressive load

4.1. Experimental setup

All small-scale TSCs were tested using a 1000 kN universal testing machine (Maekawa Testing Machine MFG Co., Ltd., Tokyo, Japan), as shown in Figure 3. Bare small-scale TSC (NS) and three small-scale TSCs strengthened with *Type A, C*, and *D* CFRP layers (namely UT0, BT0/90, and UT90) were used for compressive tests. These specimens were held in a groove fitted to a steel cylinder at a depth of 2 mm on the top and bottom ends and were sandwiched by the loading steel plate (see Figure 3).

Table 2. Models in FEA in the cases of $\sigma_h/\sigma_y = 0.5$.

Model	(NS1)	(A6-1)	(B6-1)	(C6-1)	(A12-1)	(B12-1)	(C12-1)
	Without strengthening	$(0°)_{6;0\text{-}20}$	$(\pm45°)_{6;0\text{-}20}$	$(0°/90°)_{6;0\text{-}20}$	$(0°)_{12;0\text{-}20}$	$(\pm45°)_{12;0\text{-}20}$	$(0°/90°)_{12;0\text{-}20}$
Model		(A6-2)	(B6-2)	(C6-2)	(A12-2)	(B12-2)	(C12-2)
		$(0°)_{6;0\text{-}50}$	$(\pm45°)_{6;0\text{-}50}$	$(0°/90°)_{6;0\text{-}50}$	$(0°)_{12;0\text{-}50}$	$(\pm45°)_{12;0\text{-}50}$	$(0°/90°)_{12;0\text{-}50}$
Model		(A6-3)	(B6-3)	(C6-3)	(A12-3)	(B12-3)	(C12-3)
		$(0°)_{6;0\text{-}100}$	$(\pm45°)_{6;0\text{-}100}$	$(0°/90°)_{6;0\text{-}100}$	$(0°)_{12;0\text{-}100}$	$(\pm45°)_{12;0\text{-}100}$	$(0°/90°)_{12;0\text{-}100}$

Table 3. Models in FEA in the cases of $\sigma_h/\sigma_y = 0.7$.

Model	(NS2)	(A6-4)	(B6-4)	(C6-4)	(A12-4)	(B12-4)	(C12-4)
	Without strengthening	$(0°)_{6;0\text{-}20}$	$(\pm45°)_{6;0\text{-}20}$	$(0°/90°)_{6;0\text{-}20}$	$(0°)_{12;0\text{-}20}$	$(\pm45°)_{12;0\text{-}20}$	$(0°/90°)_{12;0\text{-}20}$
Model		(A6-5)	(B6-5)	(C6-5)	(A12-5)	(B12-5)	(C12-5)
		$(0°)_{6;0\text{-}50}$	$(\pm45°)_{6;0\text{-}50}$	$(0°/90°)_{6;0\text{-}50}$	$(0°)_{12;0\text{-}50}$	$(\pm45°)_{12;0\text{-}50}$	$(0°/90°)_{12;0\text{-}50}$
Model		(A6-6)	(B6-6)	(C6-6)	(A12-6)	(B12-6)	(C12-6)
		$(0°)_{6;0\text{-}100}$	$(\pm45°)_{6;0\text{-}100}$	$(0°/90°)_{6;0\text{-}100}$	$(0°)_{12;0\text{-}100}$	$(\pm45°)_{12;0\text{-}100}$	$(0°/90°)_{12;0\text{-}100}$

4.2. Finite element analysis

Three-dimensional (3D) nonlinear analysis (LUSAS package) was used to analyze the failure buckling modes and the ultimate loads of CFRP-strengthened TSCs under compressive loads. Figure 4 shows the models of TSCs in FEA. In these models, the TSC walls were simulated using 3D quadratic 20-node higher-order solid elements (HX20). The CFRP layers were modeled as 8-node higher-order shell elements (QSL8) outside the TSCs with orthotropic materials. Half models were applied because the TSCs were symmetrical through the meridional direction. Three elements were applied in the thickness direction of TSCs. In the TSC walls, 50 and 120 elements were presented in the half-circumferential direction and in the height direction, respectively (where meshing sizes are approximately 1.66×8mm on the circumferential surfaces). The analysis iteration increments were set to the values of 0.6% maximum loading for the experiments.

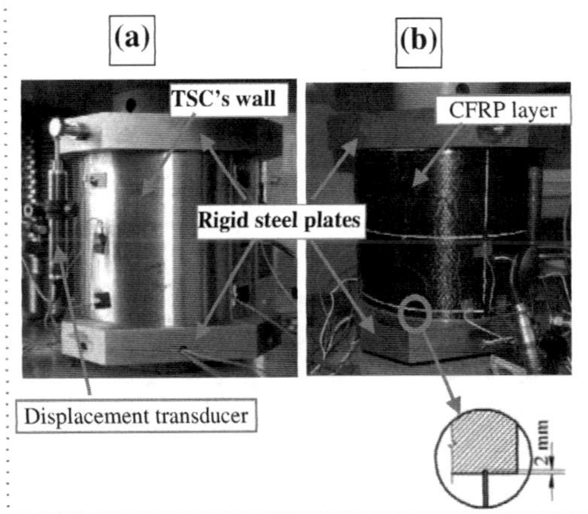

Figure. 3 Experimental setup for (a) bare TSC specimen and (b) CFRP-reinforced TSC specimens.

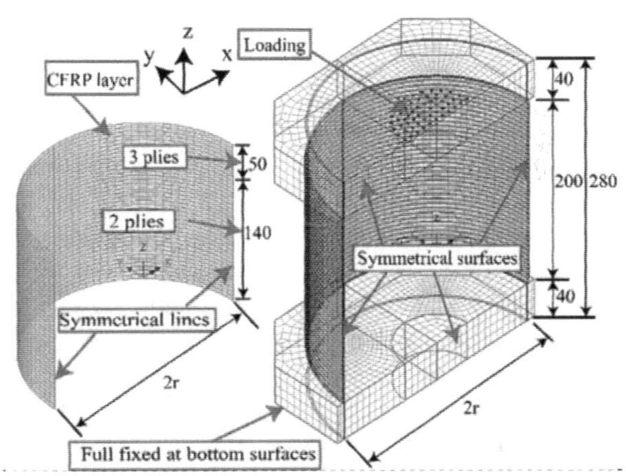

Figure. 4 Small-scale TSC models in FEA. Unit: mm.

Initial geometric imperfections were considered in FEA. The initial geometric imperfections for TSCs were assumed following equation (2) [14].

$$w = w_0 \sin(\pi y / L)\cos n\phi, \quad L = 1.728\sqrt{r_0 t_s} \qquad \textbf{(2)}$$

where y is the axial coordinate from one end of the TSC, ϕ is the circumferential angle (radian), w_0 is the amplitude of the imperfection (w_0=2%), L is the half-wavelength of the imperfection in the meridional direction, r_0 is the radius of TSC middle surface, t_s is the TSC's thickness, and n is the number of circumferential waves of the imperfection.

4.3. Experimental and FEA results

Figure 5 shows the failure modes of TSCs under compressive loads. The failure modes were different when the cylinders were strengthened by different types of CFRP layers. The failure mode was diamond buckling at the center in specimen UT0 when strengthened with circumferential CFRP layers; whereas, EFB occurred at the base of the TSC in the case of the UT90 specimen. Debonding occurred between the CFRP layer and steel for the UT90 specimen. There was a combination of EFB and diamond buckling (DB) in the case of the BT0/90 specimen. The main failure mode was EFB at the base of the cylinder in the BT0/90 specimen. However, some DB occurred at the bottom corner parts of the cylinder in the case of the BT0/90 specimen. There were good agreements between FEA and experimental results about failure modes. However, the debonding failure between the CFRP layer and steel could not perfectly be analyzed in FEA.

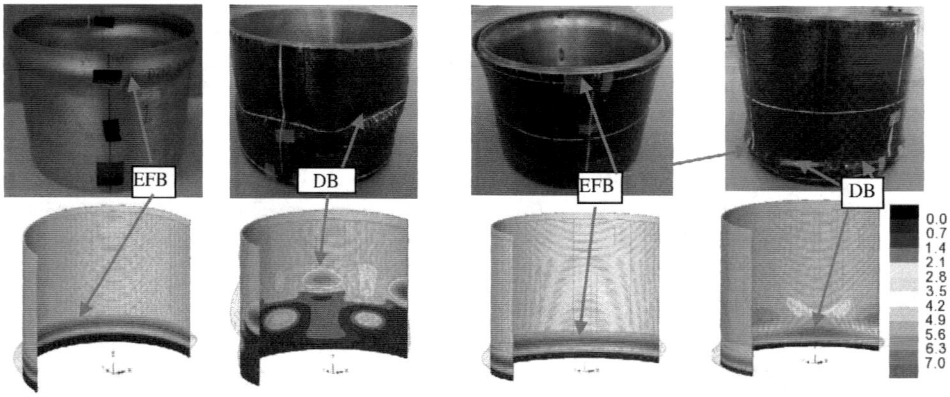

(a) NS (b) UT0 (c) UT90 (d) BT0/90

Figure. 5 Deformation of TSCs specimens in experiments and FEA. Unit (mm)

Figure 6 shows the load-displacement relations obtained from experiments and FEA. The displacements were measured from the displacement transducers, as shown in Figure 3. There were quite good correspondences between experimental and FEA results. The load-displacement relations were different after steel passing the yield stress until reaching the maximum loading. The reason is that some minor damage likely occurred between CFRP and steel after steel passing yield stress and before the breaking of CFRP layers. However, out-of-plane deformation could be confined by circumferential CFRP even if the bonding layer was damaged. So, loading was continuously increased after out-of-plane deformation reached enough value for fixing again the connection between CFRP and steel wall. Then, the maximum loads obtained from FEA were good correspondences with experimental results.

Tables 4 shows the maximum loads and strengthening effects obtained from experiments and FEA. There were good agreements between FEA and experimental results. It can be seen that the circumferential CFRP layer (UT0) had the best strengthening effect, following by the BT0/90 CFRP layer. This is because the main failure mode of the cylinder was EFB, so circumferential can be effectively constrained the buckling, leading to an increase of the load-carrying capacity. Only 90° CFRP layer had a low strengthening effect if the cylinders were subjected to axial compression.

Table 4. The maximum loads and strengthening effects of CFRP for TSCs

Specimen	Experiment				FEA			
	NS	UT0	UT90	BT0/90	NS	UT0	UT90	BT0/90
Maximum load (kN)	449.5	567.5	501.2	553.7	456.0	564.5	496.9	563.1
Strengthening effect (%)		26.3%	11.5%	23.2%		25.6%	10.6%	25.3%

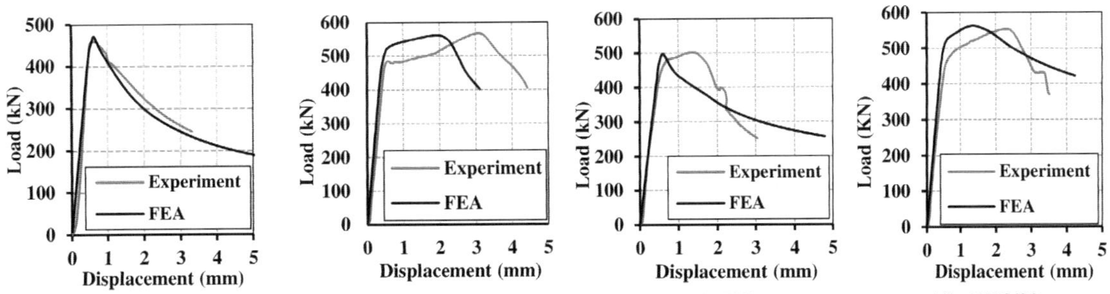

(a) NS (b) UT0 (c) UT90 (d) BT0/90

Figure. 6 Load-displacement relations of all specimens.

5. Effects of the angles of carbon fibers on the strengthening of TSCs under bending shear loads

5.1. Finite element analysis models

Figure 7 shows the FEA models for full-scale TCSs. A total of 38 models were analyzed in FEA, as shown in Tables 2 and 3. Half models were used because of symmetrical conditions. In these models, the TSC's walls were simulated with 3D quadratic solid elements and the TSC's roofs were used the surface elements with rigid materials. These rigid elements can help the load transfer from the top of the TSC to the TSC's walls easily. Mesh sizing and the number of analysis iterations play important roles in the analysis. In the TSC's walls, there were 170 elements in the circumferential direction, 80 elements in the height direction, and 2 elements in the thickness direction. The mesh sizing was dived smaller at the bottom of the TSC so that the convergence of the analysis occurred easily.

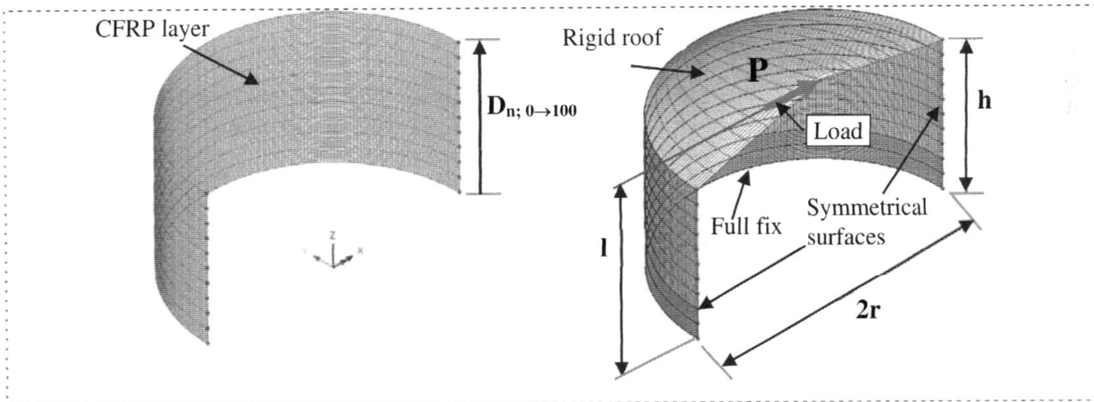

Figure. 7 Half model in finite element analysis and boundary conditions. Unit: mm.

In this study, the analysis was stopped when getting one of the two conditions. Firstly, the maximum displacement at the top of the TSC's wall was constrained to the value of one percent of the TSC's height (50mm). This is because the big displacement may affect the equipment connected to the TSCs. Secondly, in the cases of strengthening with CFRP layers, the analysis results considered the damage of CFRP layers. The analysis was stopped when the strain in the CFRP layers reached the breaking strain of CFRP layers in fiber direction (1.4%). This breaking strain value was calculated from the properties of the CFRP layer.

The loads were applied to two steps. In the first step, only the internal pressure was applied to the analysis. After finishing this step, the results were fixed and then the horizontal loading was assigned. The values of internal pressures were defined according to the ratios of σ_h/σ_y. Two levels of σ_h/σ_y (0.5 and 0.7) were surveyed. From the tensile hoop stress σ_h, the values of the internal pressure were defined by equation (1). The values of internal pressure were 0.235 (MPa) for $\sigma_h/\sigma_y = 0.5$ and 0.329 (MPa) for $\sigma_h/\sigma_y = 0.7$.

Geometrical initial imperfections were neglected in the FEA for full-scale models because these TSCs have very large dimensions and high values of internal pressures. The internal pressures create pressures that can balance the initial geometrical imperfections.

5.2. Buckling modes

Figures 8 and 9 show the buckling modes and von Mises stress of TSCs with all the cases of non-strengthened models and strengthened models. The main buckling mode of TSCs without CFRP layers is EFB. EFB usually appears at the bottom of the TSCs (20% of the height of the TSCs). The EFB occurs easily at the bottom if the ratio σ_h/σ_y is larger. When strengthening with CFRP layers, the EFB at the bottom of the TSCs was limited and the shear buckling occurred at the TSC's walls. EFB was moved upper if the numbers of CFRP layers in the circumferential direction was thick enough to restrain this buckling at the bottom as TSCs: A12-1, C12-1, A6-4, C6-4, A12-4, C12-4, A12-5, and C12-5. The prevention of EFB did not occur when using (±45°) CFRP layers. After strengthening with CFRP layers, although EFB was partly prevented, the main failures of TSCs were also EFB while more shear buckling occurred on the TSC's walls.

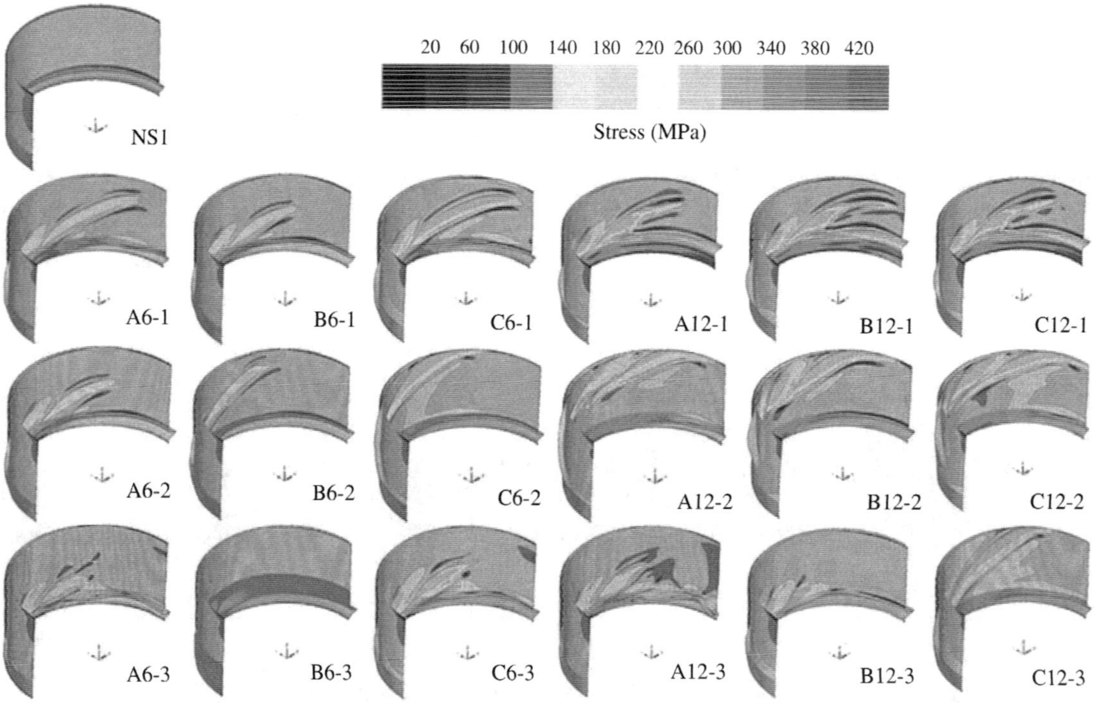

Figure. 8 The von Mises stress of TSCs in the cases of $\sigma_h/\sigma_y = 0.5$.

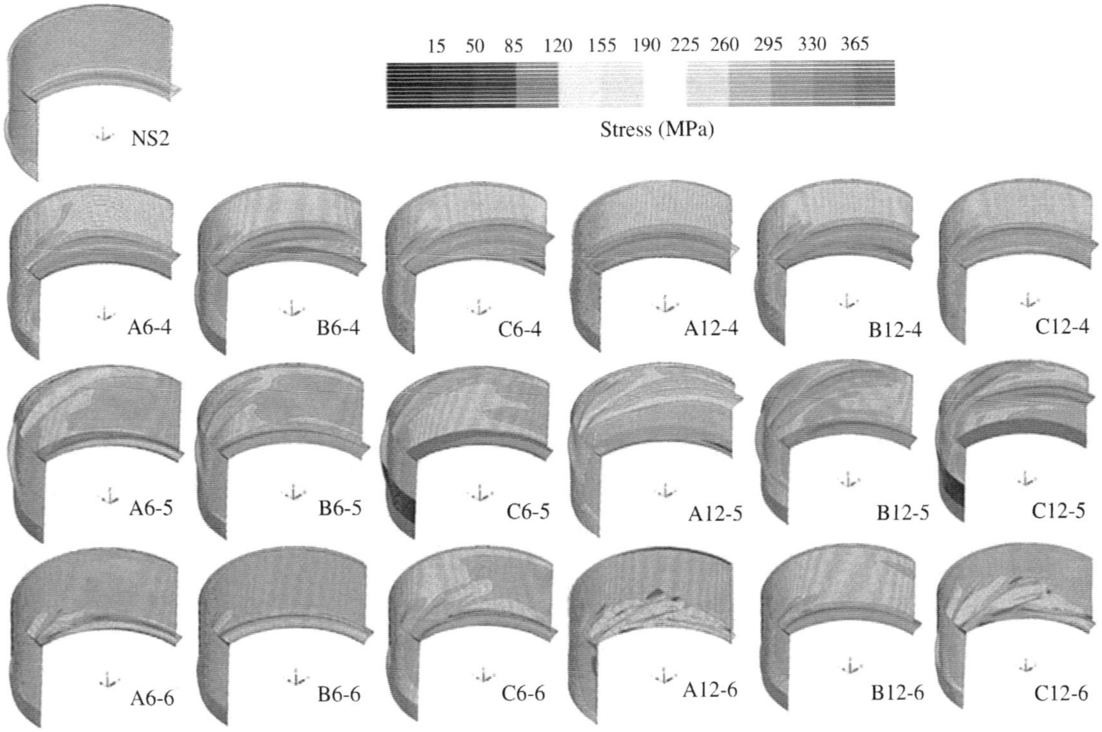

Figure. 9 The von Mises stress of TSCs in the cases of $\sigma_h/\sigma_y = 0.7$.

5.3. *The strengthening effects of CFRP layers on the increase of the load-carrying capacity of TSCs*

The maximum horizontal loads depend on the values of internal pressure, the numbers of strengthening CFRP layers, the strengthening height, and the angle of CFRP layers. Figures 10 and 11 show the maximum loads (for half models) and strengthening effects when TSCs were strengthened with 6 and 12 CFRP layers, respectively.

5.3.1. *The effects of the numbers of CFRP layers for the strengthening*

The maximum loading decreased when the values of the internal pressure increased with all cases of TSCs. When strengthening with CFRP layers, the maximum horizontal loads of strengthened models increased comparing with non-strengthened models having the same values of internal pressures. Moreover, the increase in maximum loading was corresponding to the increase in the number of CFRP layers. However, when increasing the thickness of strengthening CFRP layers to twice, the increasing effects were not double increase. For example, the increasing effect was 38% for 6-ply CFRP strengthening layer but only 61% for 12-ply CFRP layers ($\sigma_h/\sigma_y = 0.7$).

5.3.2. *The effects of the height of strengthening CFRP layers*

The results showed that the strengthening effects highly depend on the values of internal pressures. The strengthening effects were higher with higher values of internal pressures in the same strengthening heights. The strengthening effects were nearly the same with 20% and 50% strengthening heights if the ratios σ_h/σ_y equal to 0.5, whereas, the strengthening effects were a little bit changed if the ratios between hoop stress and yield stress of TSCs were increased ($\sigma_h/\sigma_y = 0.7$). From the results, it is better to choose the method of 20% of heights for strengthening because of economical effectiveness if the ratios σ_h/σ_y are small. It can be seen that the strengthening effects were highly increased when strengthening with 100% of heights in the cases of *Type A* and *C* CFRP layers, especially with higher values of the ratios σ_h/σ_y.

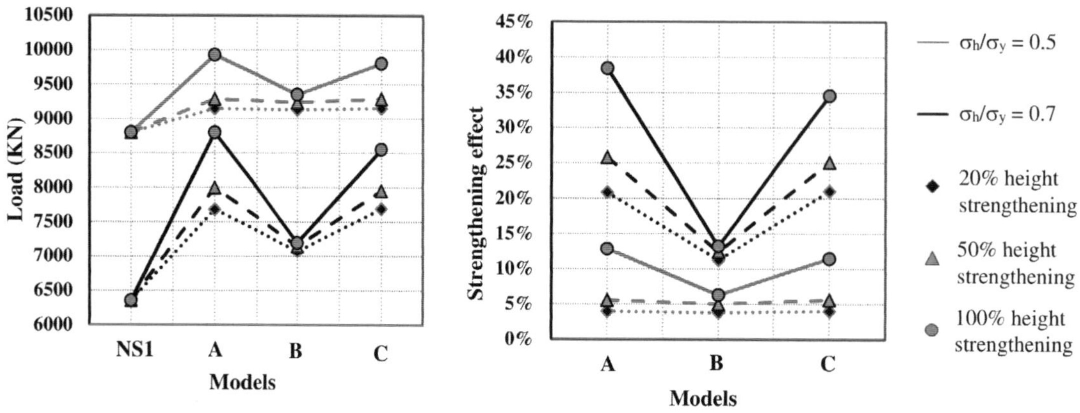

(a) Maximum loads (b) Strengthening effects

Figure. 10 Maximum loads and strengthening effects for TSCs strengthened by 6 CFRP layers.

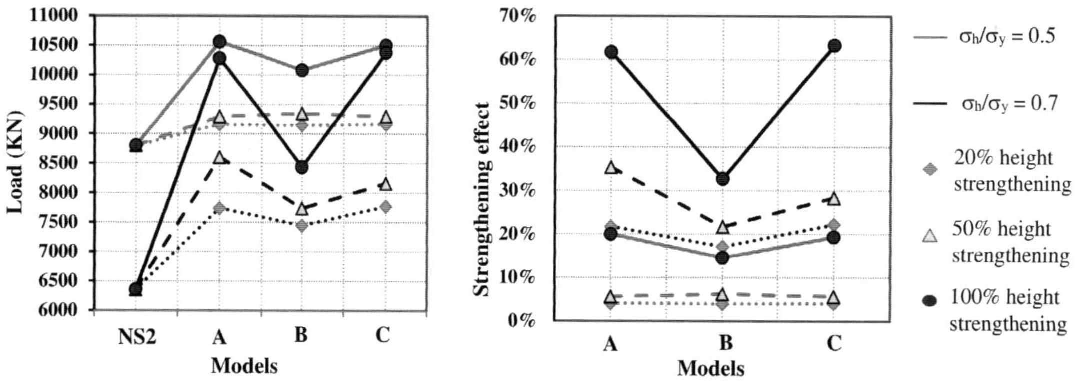

(a) Maximum loads (b) Strengthening effects

Figure. 11 Maximum loads and strengthening effects for TSCs strengthened by 12 CFRP layers.

5.3.3. The effects of the angles of fibers of strengthening CFRP layers

The strengthening effects of different types of CFRP-strengthened layers are related closely to the buckling restraint of the TSCs. The unidirectional CFRP layers can restrain the EFB effectively; whereas the angle-CFRP layers have significant impacts on SB restraint. The failure modes were EFB in the TSCs without strengthening. When strengthening with CFRP layers, SB occurred on the TSC's walls but the dominant modes were also EFB. This is because the high values of σ_h/σ_y were focused on this study. That is the reason *Type A* (0°) and *C* (0°/90°) CFRP layers had high strengthening effects for the TSCs. From the results, it is better to choose the circumferential (0°) or (0°/90°) CFRP layers to strengthen for TSCs that are subjected to high values of internal pressures under bending shear loads.

6. Conclusion

In this study, the effects of the angles of carbon fibers for the strengthening of TSCs under compressive loads and bending shear loads were investigated. Some following main points are highlighted.

- Circumferential CFRP layers revealed their significant effects on the increased load-carrying capacity of TSCs under axial compression. The appearance of circumferential CFRP layers would bring higher strengthening effects for buckling restraint of TSCs under axial compression. Only 90 degrees of CFRP layers should not be used to strengthen the TSCs under axial compression.

- The main buckling failure of full-scale TSCs without CFRP strengthening was EFB. When strengthening with CFRP layers, the EFB at the bottom of the TSCs was limited and the SB occurred at the TSC's walls.
- The EFB at the bottom of the TSCs could be prevented when using enough CFRP layers at the bottom of the TSCs (20% of heights) with *Type A* and *C* CFRP layers. The EFB moved upward if the thickness of the CFRP layer was enough to prevent the EFB.
- When TSCs were subjected to internal pressure with the ratios $\sigma_h/\sigma_y = 0.5$, the strengthening effects were nearly the same with 20% and 50% strengthening heights. Therefore, it is better to choose 20% strengthening heights because of economic benefit.
- *Type A* and *C* had better strengthening effects than *Type B* CFRP layers for TSCs under bending shear loads because the main failure modes of TSCs were EFB. It is recommended that only angled fiber of CFRP layers should not be used to strengthen when the TSCs are subjected to high values of internal pressures and bending shear loads.

References

[1] Architectural Institute of Japan (AIJ) 2012 *Reconnaissance Report On The 2010 Chile Off Maule Earthquake and Reconnaissance Report On The 2011 New Zealand Christchurch Earthquake* Japan.
[2] Architectural Institute of Japan (AIJ) 2012 *Preliminary Reconnaissance Report of The 2011* Tohoku-Chiho Taiheiyo-Oki Earthquake Japan.
[3] Zhao X L and Zhang L 2007 State-of-the-art review on FRP strengthened steel structures *Engineering Structures* **29** 1808–23
[4] Harries K A, Peck A J and Abraham E J 2009 Enhancing stability of structural steel sections using FRP *Thin-Walled Structures* **47** 1092–101
[5] Neale K W 2000 FRP for structural rehabilitation: A survey of recent progress *Progress in Structural Engineering and Materials* **2** 133–38
[6] Cadei J M C, Stratford T J, Hollaway L C and Duckett W H 2004 *C595-Strengthening Metallic Structures Using Externally Bonded Fiber-reinforced Composites* London: CIRIA
[7] Teng J G, Yu T and Fernando D 2012 Strengthening of steel structures with fiber-reinforced polymer composites *Journal of Constructional Steel Research* **78** 131-43
[8] Batikha M, Chen J F, Rotter J M and Teng J G 2009 Strengthening metallic cylindrical shells against elephant's foot buckling with FRP *Thin-Walled Structures* **47** 1078–91
[9] Bhetwal K K and Yamada S 2012 Effects of CFRP reinforcements on the buckling behavior of thin-walled steel cylinders under compression *International Journal of Structural Stability and Dynamics* **12** 131–51
[10] Nhut P V and Matsumoto Y 2018 The Effects of Carbon Fiber Reinforced Polymer Strengthening on Cylindrical Steel Storage Tanks under Bending Shear Load *IOP Conf. Series: Materials Science and Engineering* **371** 012025
[11] Nhut P V, Matsumoto Y, Matsui T and Nakamura H 2019 Strengthening effects of circumferential CFRP layers for thin-walled steel cylinders under axial compression *Steel Construction Engineering* **26** 15-29
[12] Nhut P V, Kitano Y and Matsumoto Y 2020 Experimental Investigations of the Strengthening Effects of CFRP for Thin-Walled Storage Tanks under Dynamic Loads *Applied Sciences* **10** 2521
[13] Nhut P V and Matsumoto Y 2020 Experimental analytical and theoretical investigations of CFRP strengthened thin-walled steel plates under shear loads *Thin-Walled Structures* **155** 106908
[14] Teng J G, Hu Y M 2007 Behaviour of FRP-jacketed circular steel tubes and cylindrical shells under axial compression *Construction and Building Materials* **21** 827–88

The dissolution of hematite deposits on model spent ion exchange resins using direct current

E A Tokar[1], A I Matskevich[1], T A Sokolnitskaya[1], A M Egorin[1]

[1]Institute of Chemistry, Far Eastern Branch, Russian Academy of Sciences, prospect 100-letiya Vladivostoka, 159, 690022, Vladivostok, Russia

E-mail: andrey.egorin@gmail.com

Abstract. The method of dissolving hematite deposits intensification, containing cobalt radionuclides, on ion exchangers KU 2-8, modelling spent ion exchange resin, was proposed. Defined, that with direct current increasing and with acids concentration increasing, the efficiency of hematite deposits dissolution is increasing and could exceed 99%. The hematite dissolution accompanied by radionuclides secondary adsorption due to ion exchange that could be removed with sodium nitrate solution was presented.

1. Introduction

Because of partial corrosion of steel elements, the oxide/hydroxide particles were formed in condensation water, contour NPP [1, 2]. This particles have the common name – crud, its chemical composition is mainly iron and less chrome, cobalt, nickel and zinc [3, 4]. The crud removing proceeded with mechanical filtration on ion exchange resins, with subsequent backwash. The increasing of ion exchange resins exploitation time leads to irreversible pollution, because of facilitated diffusion crud particles to ion exchangers pore [5]. This particles cannot be removed with subsequent backwash, and the efficiency of its removing, using chemical regeneration, does not exceed 50% [5]. The decontamination of spent ion exchange resins (SIER) is a complex technological problem, because the crud particles inclined to short-lived radionuclides adsorption of the corrosive group [3]. Therefore, the treatment SIER with salts and acids solutions without the iron oxide deposits dissolution, is not effective. The problem of SIER decontamination complicates by the fact that the efficiency of iron oxide deposits removing will be determined by complete hematite dissolution, that appeared during magnetite oxidation dissolved oxygen [2] or because of goethite transformations [1]. The hematite is hardly dissolved in mineral acids compared with magnetite [6] and wherein practically is not removed from SIER with ultrasonic treatment.

The well-known methods of recycling or treatment SIER as immobilization to cement matrices, pyrolysis, combustion, have some technical disadvantages. This fact forces researchers to look for a new approaches to safer and more perfect methods of recycling SIER. The methods of liquid phase oxidation using Fenton and electro-Fenton process [7, 8] are considered quite promising. However, the liquid phase oxidation disadvantages are using concentrated solution of hydrogen peroxide and incompleteness of resins dissolution.

There is not these disadvantages in the non-destructive methods of SIER decontamination. These methods are effective at silicate deposits dissolution and was presented by the authors of this article in work Palamarchuk et al. [9]. One of this is method of electrochemical regeneration SIER [10, 11]. This method assumes the treatment SIER in citrates solutions and ascorbic acid, accompanied by partial iron reduction with transition to a dissolve form, in form of complex compounds. Unfortunately, the efficiency of SIER decontamination with this method does not exceed 40%. Undoubtedly, the reductive hematite dissolution is perspective method of SIER decontamination from iron oxide deposits, besides, adding organic compounds makes some difficulties with cleaning the appeared secondary radioactive wastes. In our opinion, the optimal method resins decontamination, which polluted by iron oxide deposits, is treatment in near-cathode region in mineral acids solutions. Therefore, the purpose of this work is assessment of practical application the SIER decontamination under the electric current.

2. Experimental

SIER were received by modified method, presented in work [12]. The ion exchange resin KU 2-8 was transferred to Fe-form using the $FeCl_3$ solution (0.5 mol/L) marked by Co-57 radionuclide. Then, the NH_4OH solution was added to the ion exchange resin and with hydrothermal treatment within 24 hours at 175 °C temperature and pH 2. After hydrothermal treatment, the SIER samples were washed by solution with composition: HNO_3 - 3 mol/L, $NaNO_3$ – 1 mol/L. The hydrothermal treatment leads to forming of hematite phase in volume of resin grains (Figure 1).

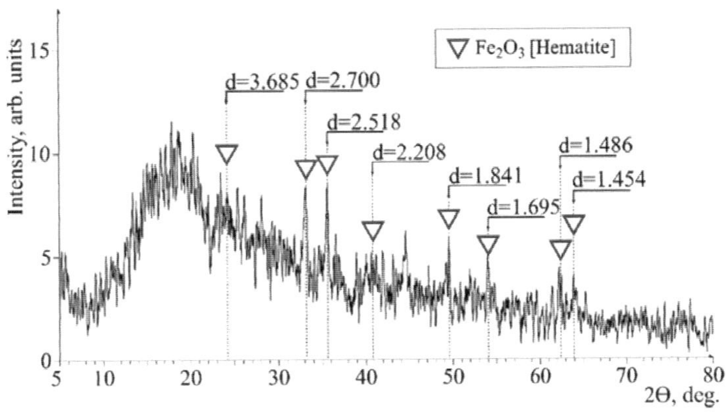

Figure. 1 XRD patterns of model SIER

The treatment of model SIER was made using electrochemical cell, the construction is presented on Figure 2.

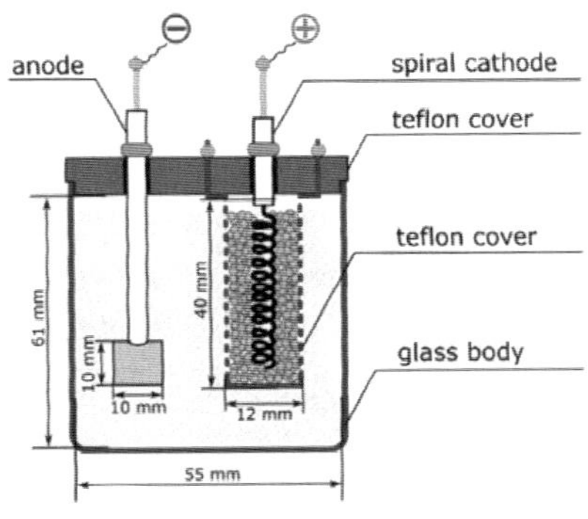

Figure. 2 The construction of electrochemical cell.

Into the chamber of an electrochemical cell with volume 100 mL, the electrodes and the basket for model SIER retention in near-cathode region were set. The basket's diameter is 12 cm, the working volume is 3.6 ml. As a cathode material we used a platinum coiled wire with purity 99.99%, the diameter is 0.5 mm and the length is 150 mm. The treatment were made using the constant current source in current stabilization mode. To evaluate the iron oxide deposits dissolution through the certain time intervals, the fixed volume of solution was selected (1-2 ml), where the concentration of iron was determined. The spent part of solution was replaced by fresh acid solution, further we considered it in the calculations. For assessment of the SIER decontamination efficiency through the certain time intervals, the solution, with volume 10 ml was selected and the specific activity was measured. After the measurement, the solution was returned to electrochemical cell.

3. Results and discussion

On Figure 3 the diagram of efficiency model SIER decontamination with sulfuric and nitric acid solution adding $NaNO_3$ (3 mol/L), decreasing the radionuclide secondary adsorption, was presented. Using the sulfuric acid solution, with concentration 1 mol/L, the efficiency of model SIER decontamination achieves to 80 %, wherein the nitric acid, at the same conditions, is not effective for hematite dissolution. Therefore, using the mineral acids solutions does not provide the complete SIER decontamination, because the hematite is hardly dissolve, and it is necessary to use additional methods.

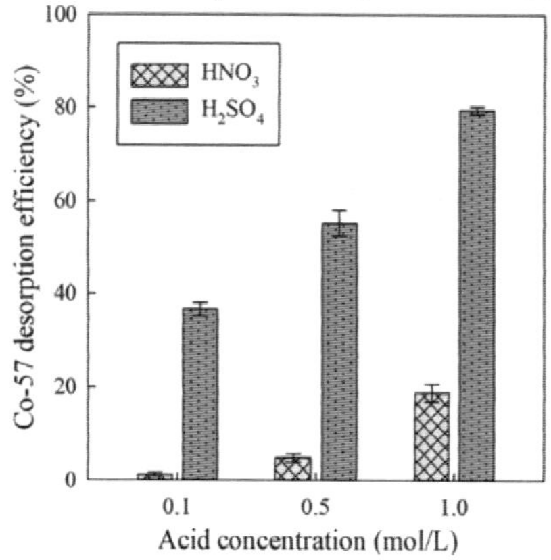

Figure. 3 The diagram of dependence between the model SIER decontamination type and acid solution concentration

The additional SIER treatment, with direct current, could increase the efficiency of hematite dissolution and depends on reactions below:

1. $2H^+ + 2ē \rightarrow 2H_2\uparrow$ (cathod)
2. $2H_2O - 4ē \rightarrow O_2\uparrow + 4H^+$ (anode)
3. $Fe_2O_3 + 6H^+ + 2ē \rightarrow 2Fe^{2+}_{(aq)} + 3H_2O$ [13].

Thus, in cathode space there is the hematite dissolve intensification, due to recovery processes. There are the kinetic curves of hematite dissolution on Figure 4. To assess the impact of sulfuric acid concentration and current strength to the hematite dissolution speed, the initial sections of kinetic curves were approximated by linear functions (Figure 4a), the parameters are on Table 1.

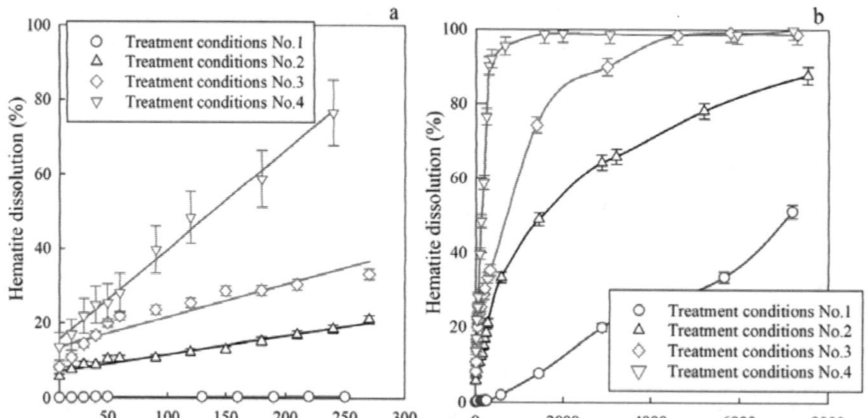

Figure 4. The kinetic curves of iron oxide deposits dissolution on model SIER (a) and initial sections of kinetic curves (b)

We can note, that increasing of current strength on 0.5 A, with fix sulfuric acid concentration – 1 mol/L, leads to doubled increasing of dissolution speed (mode No. 2 and 3, parameter *a*). The increasing of sulfuric acid concentration from 1 mol/L to 2 mol/L, and also current strength, leads to threefold increasing of dissolution speed (treatment No. 3 and 4). Thus, the hematite dissolution speed in cathode space is determined not only by sulfuric acid concentration, and also by current strength.

Table 1. The parameters of linear functions, received at initial sections approximation of kinetic curves

No.	Treatment conditions			Linear equation parameters		
	Concentration H_2SO_4 (mol/L)	Current strength (A)	Voltage (V)	*a*	*b*	R^2
1	0.1	0.5	9.9	0.0013	0.1963	0.6644
2	1.0	0.5	3.0	0.0495	6.6893	0.9663
3	1.0	1.0	3.7	0.0886	12.7496	0.8600
4	2.0	2.0	5.0	0.2676	12.9308	0.9908

In Table 2, there are the results of dependence the efficiency total iron oxide deposits dissolution on type and resin treatment conditions. Using the electrochemical treatment, allows to achieve highly efficiency of hematite deposits dissolution on model SIER, 99.3 and 99.7% for modes No. 3 and 4, respectively. Based on received results, the further work was carried out using treatment conditions No. 3.

Table 2. The efficiency of hematite dissolution on SIER

Treatment conditions (No.)	Efficiency of dissolution (%)
1	48.7
2	82.2
3	99.3
4	99.7

In Figure 5, there is the curves of dependence between model SIER decontamination and electrochemical treatment time. Because of KU 2-8 is ion exchange resin, its property does not change in hydrothermal treatment process [12], and there is the radionuclide Co-57 fixation on iron oxide deposits (fixed part) and on functional resin sulfo-group (exchange part). For this reason, the SIER decontamination curve passes through the maximum, which indicates the process of radionuclide Co-57 secondary adsorption, entered into the H_2SO_4 solution, due to hematite dissolution, on sorption ion exchanger centers. For adsorbed radionuclide removing, ion exchangers after electrochemical treatment, were additionally washed by solution: $NaNO_3$ – 2 mol/L, HNO_3 – 1 mol/L. As a result of this washing, Co-57 photopeaks were not detected, which indicates complete the model SIER decontamination.

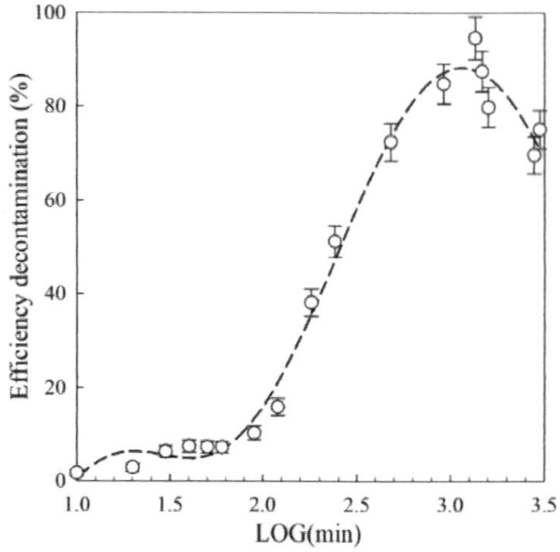

Figure 5. The dependence between decontamination SIER efficiency and direct current treatment time.

For decreasing the secondary wastes, the solutions, after model ion exchangers decontamination, were alkalized using NaOH. Changing pH solution leads to the precipitation of $Fe(OH)_3$ with the simultaneous the Co-57 radionuclide coprecipitation. After solution centrifugation, the Co-57 photopeaks were not detected, which indicates the complete radionuclide coprecipitation on the $Fe(OH)_3$ precipitate whose wet weight was 0.2 g.

4. Conclusion

The method of hematite deposits dissolve intensification on ion exchange resins, with direct electric current treatment, was proposed and tested. Increasing concentration of H_2SO_4 solution, and also direct current, the speed and completeness of hematite deposits dissolution, and, respectively, the efficiency of decontamination are also increasing. We showed, that this method can quickly and efficiently dissolve hematite particles, compared with usual acid treatment. Adding the reductants, such as oxalic and ascorbic acids, does not need, this is the advantage of the method, we used. It was showed, that the process of model SIER decontamination accompanied by gradual hematite dissolution with subsequent Co-57 secondary adsorption on the functional ion exchangers groups, therefore the additional treatment with sodium salts for radionuclide desorption.

Acknowledgments

The authors gratefully acknowledge the Russian Science Foundation for supporting this work (Project 18-73-10066). Equipment of CUC “Far Eastern Center of Structural Investigations” was used in this work.

Reference

[1] Sawicki J A 2011 Analyses of fuel crud and coolant-borne corrosion products in normal water chemistry BWRs *Journal of Nuclear Materials* **419** 85–96
[2] Sawicki J A, Sefranek P J and Fisher S 1998 Depth distribution and chemical form of iron in low cross-linked crud-removing resin beds *Nuclear Instruments and Methods in Physics Research Section B: Beam Interactions with Materials and Atoms* **142** 122–32
[3] Hoshi M, Tachikawa E, Suwa T, Sagawa C, Yonezawa C, Aoyama I and Yamamoto K 1986 Crud Behaviors in High-Temperature Water, (I): Characterization of Water in JMTR OWL-1 Loop *Journal*

of Nuclear Science and Technology **23** 511–21
[4] Tsai T-L, Lin T-Y, Su T-Y, Wei H-J, Men L-C and Wen T-J 2012 Identification of chemical composition of CRUD depositing on fuel surface of a boiling water reactor (BWR-6) plant *Energy Procedia* **14** 867–72
[5] Otoha K, Izumi T, Hayashi T, Morikawa Y and Murabayashi H 1996 Crud Removal Performance with Ion Exchange Resins in BWR Plants *Journal of nuclear science and technology* **33** 10
[6] Sidhu P S 1981 Dissolution of Iron Oxides and Oxyhydroxides in Hydrochloric and Perchloric Acids *Clays and Clay Minerals* **29** 269–76
[7] Xu L, Meng X, Li M, Li W, Sui Z, Wang J and Yang J 2019 Dissolution and degradation of nuclear grade cationic exchange resin by Fenton oxidation combining experimental results and DFT calculations *Chemical Engineering Journal* **361** 1511–23
[8] Cheng T-H, Huang C-P, Huang Y-H and Shih Y-J 2017 Kinetic study and optimization of electro-Fenton process for dissolution and mineralization of ion exchange resins *Chemical Engineering Journal* **308** 954–62
[9] Palamarchuk M, Egorin A, Tokar E, Tutov M, Marinin D and Avramenko V 2017 Decontamination of spent ion-exchangers contaminated with cesium radionuclides using resorcinol-formaldehyde resins *Journal of Hazardous Materials* **321** 326–34
[10] Chi L and Semmler J 2010 Electrochemical regeneration of spent ion exchange resin (Atomic Energy of Canada Limited) *Nuclear Plant Chemistry Conference (NPC2010) Report AECL-CW--127140-CONF-002*
[11] Semmler J and Chi L 2012 Treatment of liquid waste and regeneration of spent ion exchange resin using electrochemical techniques (Atomic Energy of Canada Limited) *Conference Paper Report AECL-CW--127140-CONF-003*
[12] Egorin A, Tokar E, Kalashnikova A, Sokolnitskaya T, Tkachenko I, Matskevich A, Filatov E and Zemskova L 2020 Synthesis and Sorption Properties towards Sr-90 of Composite Sorbents Based on Magnetite and Hematite *Materials* **13** 1189
[13] Yanina S V and Rosso K M 2008 Linked Reactivity at Mineral-Water Interfaces Through Bulk Crystal Conduction *Science* **320** 218–22

IOP Conf. Series: Materials Science and Engineering 1048 (2021) 012021 doi:10.1088/1757-899X/1048/1/012021

Optimization of structural elements in highly seismic areas using neural networks

V. Arana[1], M. Sanchez[2] and P. Vidal[2]

[1]Professor, Peruvian University of Applied Science UPC, Lima Peru
[2]Bachelor, Peruvian University of Applied Science UPC, Lima Peru

E-mail: pccivara@upc.edu.pe

Abstract. The aim of this research is to use Artificial Neural Networks (ANN) to dimension structural elements in regular 6-storey buildings. The necessary data for the training of the algorithm was elaborated manually with the help of the ETABS software, these were 30 buildings of reinforced concrete with a system of structural walls. The configuration and training of the neural network was carried out in the MATLAB software. The validation was carried out in an additional analyzed building in which the concrete savings were calculated, and the requirements of the current regulations were verified. Finally, the dimensioning obtained with the neural network generated a reduction of more than 10% in the total volume of concrete used in a 6-level building and establishes that the algorithm used provides effective results for an optimal design.

1. Introduction

Nowadays the importance of pre-dimensioning the structural elements and not generating cost overruns is increasing. Because construction costs depend on the amount of materials (concrete, steel and formwork), i.e. the direct costs of the project, and these in turn are related to the dimensions of the sections of the structural elements. However, in order to carry out an adequate pre-dimensioning that does not generate a cost overrun of 28% of the total estimated cost [1], it is necessary to carry out an iterative process based on the designer's criteria.

In the calculation of the optimal dimensions of columns, beams and shear walls there are previous investigations in which methodologies such as artificial intelligence, quadratic sequential programming [2], among others, are used. In AI some of the most common optimization techniques are particle swarm optimization (PSO), simulated annealing (SA) and artificial neural networks.

In the context of artificial intelligence, the particle swarm optimization (PSO) technique was proposed in 2013 for the calculation of beam sections to reduce construction costs [3]. Another method is the artificial neural network, which was used to optimize structural elements (beams and columns) in buildings [4]. In these investigations, the optimization of the sections provided a saving of 6.7% and 8.4% when compared with a conventional design method [3] and with the application of the Simulated Annealing (SA) technique (explaining the process of the technique) in two building models, they achieved a saving of 33.8% and 34.1% for each one of the models [5]. However, the optimization of sections was done individually (beams or columns) [2], [3], [5]-[9][10] or in the case of a building only columns and beams.

In this research, a methodology capable of optimizing the dimensions in columns, beams and reinforced concrete shear walls is proposed, based on preliminary design data of proposed multi-family buildings. This will reduce the cost overrun of the materials that are used when the structural elements are over-dimensioned, and in turn will reduce the time to obtain the dimensions. For this purpose, a neural network algorithm will be developed to reduce by 10% the cost overrun when estimating the optimal dimensions of the structural elements (columns, beams and shear walls).

2. Methodology

To optimize the dimensions of the structural elements, neural networks were used, so buildings were modeled with a system of structural and regular 6-story walls. The preliminary dimensions of the buildings were selected as input data and the measurements of beams, columns and shear walls as output variables. Then, the training was carried out with the information collected and the validation of the neural network configured through the modeling of a new building. Finally, it was calculated the concrete difference that is obtained when dimensioning a building with neural networks and in a conventional way [11], from this, the concrete saving was calculated.

2.1. Information gathering and classification of structures

Regular structures were modeled and analyzed in the ETABS software with a minimum floor area of 100 m2 and a maximum of 1300 m2, with a story height of 3.0 m. The structural analysis was carried out using the National Building Regulation (RNE) in its chapters E 020, E030 and E060, for this purpose 30 types of structures in plant were proposed based on the indicated parameters. Then it was defined the possible dimensions that the buildings will have in plant: width (A) and length (L) and it was verified that the area that occupies is within the established limits.

The scheme of one of the cases is shown in Figure. 1, the number of columns in each axis (X and Y) were variable and at their own discretion. The location of the shear walls in both directions was strategic at the discretion of the researcher. In addition, each model has at least one elevator, located in the center of the building and in some cases at the ends, depending on the floor area. Figure. 2 shows the isometric view.

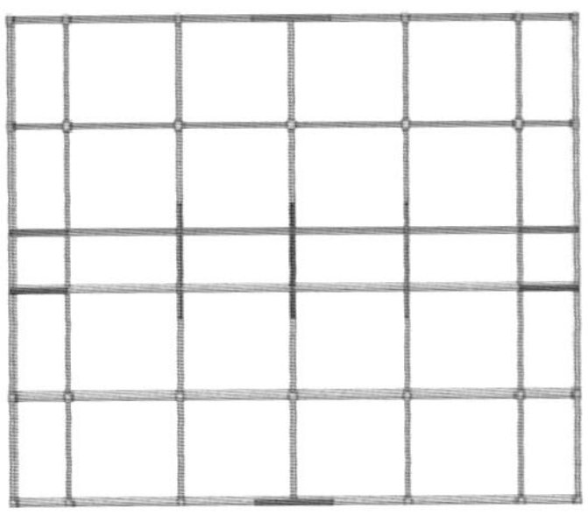

Figure. 1 Plan view type 19 (25x30m)

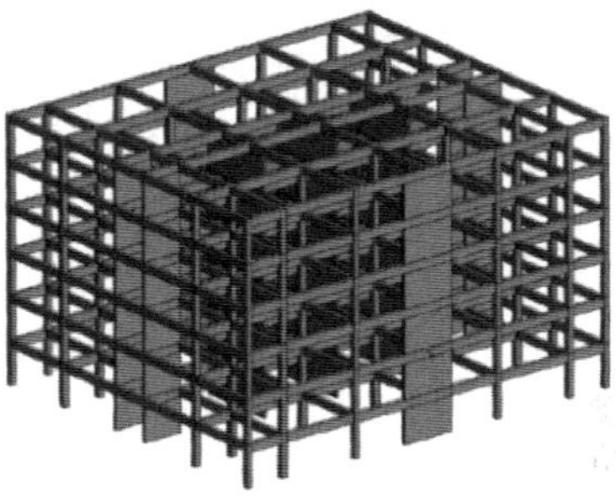

Figure. 2 Isometric view of plant type 19

For the modeling, the loads used were dead and live loads; for the seismic analysis, the static and modal spectral analysis was used with the following seismic parameters in a high seismicity zone:

- Maximum horizontal acceleration (Z) = 0.45g
- Importance Factor (U) = 1
- Soil factor (S) = 1
- Reduction Coefficient (R) = 6

Then we proceeded with the reduction of the length of the cut walls because this is more important in the lateral displacements, with that length the structural analysis was made to verify the drifts, this process was iterative and we obtained a final length of shear wall in the X and Y direction with drifts less or equal to 0.007 according to the NTP E. 0.30 (Peruvian Technical Standard). This process was carried out for all 30 models and Figure. 3 shows the comparison of drifts between the initial model which consists of the building pre-dimensioned in a conventional way and the optimized structure with the dimensions of reduced structural elements of case 19.

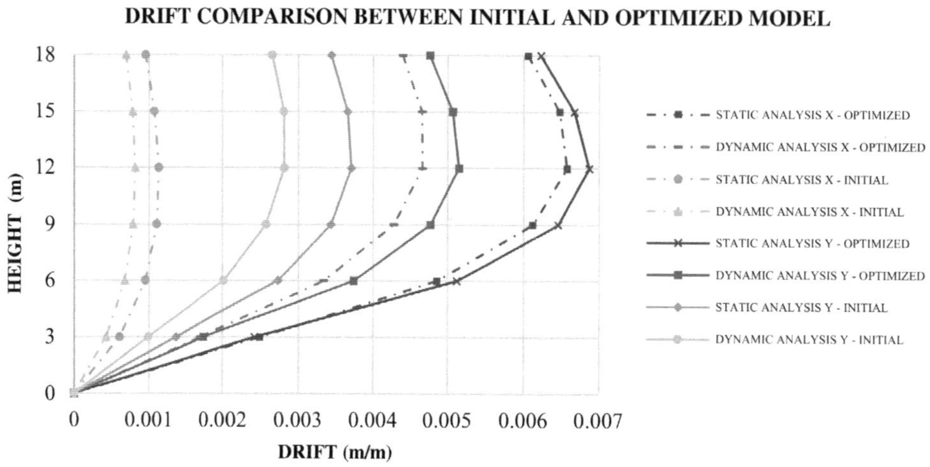

Figure. 3 Drift comparison between the initial and optimized models

It is observed that, as the length of the shear walls in both directions and in some cases the cross-sectional area of the columns was reduced, the drifts of each level gradually increased until reaching the limit of 0.007 provided by RNE E.060.After performing the indicated procedure, the dimensions shown in

Table 1 were obtained.

Table 1. Final dimensions of the structural elements

Element	Dimension	Measure (m)
Perimeter column	Width	0.40
	Depth	0.40
Central column	Width	0.45
	Depth	0.45
Beam in X direction	Width	0.30
	Depth	0.55
Beam in Y direction	Width	0.30
	Depth	0.60
Shear walls in X direction	Thickness	0.20
	Length	18.00
Shear walls in Y direction	Thickness	0.25
	Length	22.00

These dimensions were obtained after an iterative process and are those necessary to obtain a maximum drift of 0.0068 very close to the limit of 0.007, however, to be used as training data will be changed from units and columns are entered as area, then shows the data finally used in this case for training.

- Perimeter column area = 1600 cm2
- Center column area = 2025 cm2
- Length of shear wall in X direction = 1800 cm
- Thickness of shear wall in length X = 20 cm
- Length of shear wall in Y direction = 2200 cm
- Thickness of shear wall in Y direction = 25 cm
- Width of beam in X direction = 30 cm
- Depth of beam in X direction = 55 cm
- Width of beam in Y direction = 30 cm
- Depth of beam in Y direction = 60 cm

We changed from units to centimeters to obtain results with greater precision, since we work with larger numbers and in the case of columns we decided to train with the cross-sectional area because most cases use square section columns.

2.2. Architecture of the artificial neural network

For the elaboration of the neural network, input data was established, defined by the preliminary data we have of the building's architecture, and the dimensions of the structural elements will be identified as output variables as shown in Figure. 4..

Where variables A, B, C and D are the input data of the ANN width, length, maximum span in the X direction and maximum span in the Y direction; and variables 1 to 10 represent the output data representing the area of the columns located at the corners and perimeters, the area of the columns located in the internal part of the building, the accumulated length and width of the cutting walls used (X direction and Y direction), the base and height of the most critical beam in the X and Y direction.

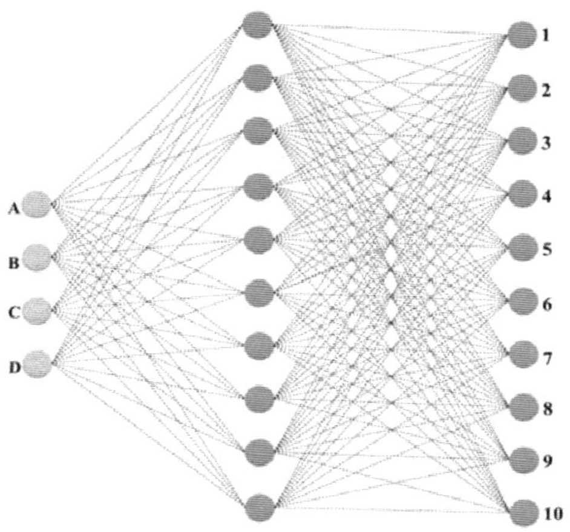

Figure. 4 Architecture of the artificial neural network

The elaboration of the neural network was carried out in the MATLAB software with the help of the "nntool". The number of neurons in the hidden layer of the neural network will be defined according to the convergence and training needs that arise in each model. A total of 10 tests were performed with different layers and numbers of neurons in each one; and the network 10 was selected for presenting greater homogeneity and correlation with respect to the other networks as shown in Table 2.

Table 2. Correlation coefficient of test networks

Net	[Input - Number of neurons - Output]	R - Training	R - Test	R - Validation	R - total
1	[4 10 10]	0.984	0.953	0.910	0.967
2	[4 15 10]	0.902	0.909	0.971	0.917
3	[4 20 10]	0.956	0.925	0.980	0.956
4	[4 10 10 10]	0.718	0.784	0.882	0.754
5	[4 10 15 10]	0.928	0.912	0.970	0.934
6	[4 15 15 10]	0.759	0.831	0.810	0.760
7	[4 20 25 10]	0.994	0.983	0.971	0.989
8	[4 20 20 10]	0.792	0.817	0.756	0.788
9	[4 25 20 10]	0.983	0.998	0.941	0.975
10	[4 25 25 10]	0.982	0.980	0.983	0.982

The database for the training of the artificial neural network was compiled by designing 30 types of buildings with the same number of floors but with column separations in both variable axes, which were structured, analyzed and designed. When the initially proposed sections had a resistance well above the requested ones, the cross sections were reduced and redesigned, until obtaining the necessary sections to satisfy the requests. The final sections obtained were used for training and validation of the neural network.

2.3. Methodology validation

In this stage the algorithm was executed with the preliminary data of a building called input variables, this building was different from those used in the training, from these data the dimensions of the structural elements were calculated. The results obtained from the analysis are shown in the following chapter.

Additionally, the neural network was executed with one of the data used in the training to validate that the results it provides are approximate, the output dimensions and the variation that was obtained in comparison to the calculated data is shown in Table 3.

Table 3. Neural network execution result

Output variable	Output ETABS	Neural network output	Variation%
Perimeter column area (cm2)	900	880.17	2.2%
Center column area (cm2)	1225	1161.96	5.1%
Shear wall length in X direction (minor) (cm)	2200	2173.75	1.2%
Shear wall thickness in X direction (minor) (cm)	20	23.77	18.9%
Length of shear walls in Y direction (major) (cm)	1900	1770.34	6.8%
Shear wall thickness in Y direction (major) (cm)	20	20.28	1.4%
Beam width in X direction (minor) X (cm)	25	27.02	8.1%
Beam depth in X direction (minor) (cm)	40	45.16	12.9%
Beam width in Y direction (major) (cm)	25	29.95	19.8%
Beam depth in Y direction (major) (cm)	50	46.74	6.5%

The maximum percentage variation obtained is 19.8% while the minimum was 1.2%, however, the maximum value was obtained in the dimension of beam base width, in which it was obtained a length of 29.95 cm by the execution of the neural network compared to 25 cm calculated, the difference was 4.95 cm being this difference acceptable in engineering.

2.4. Calculation of concrete savings

The final stage of the research consisted in comparing the volume of concrete needed in a building dimensioned with neural networks and conventionally. From this information, the difference of concrete between both structures was calculated to finally indicate the savings obtained, the results obtained are shown in the following chapter.

3. Resulted

3.1. Neural network testing

The validation of the algorithm was done by running the neural network with a new building and its input data are shown below:

- Building width of 20 m
- Building length of 21 m
- Beam span in the direction of the building width of 4.50 m
- Span of beams in direction of building length of 5.00 m

With the input data the selected neural network was executed, with this the optimal dimensions that each structural element should have were obtained, these are shown in Table 4.

Table 4. Output variables

Output variables	Test 2
Perimeter column area (cm^2)	917.17
Center column area (cm^2)	1142.69
Length of shear walls in X direction (minor) (cm)	2109.33
Thickness of shear walls in X direction (minor) (cm)	23.05
Length of Y-direction shear walls (largest) (cm)	1605.49
Thickness of Y-direction shear walls (largest) (cm)	20.14
Width of beam in X direction (minor side) (cm)	26.65
Depth of beam in X direction (minor side) (cm)	44.32
Width of Y-beam (largest side) (cm)	29.92
Depth of Y-beam (largest side) (cm)	50.38

With the dimensions provided by the neural network for the beams, columns and shear walls, the modeling and structural analysis of the building was performed, the cross section chosen for the column was square, the dimensions of the beam were rounded to a multiple of 5cm and the length of the shear walls were strategically distributed in both directions. From this, the elastic drifts were obtained and multiplied by 0.75R to obtain the inelastic drifts as indicated by RNE E.060. These results are shown in Table 5.

Table 5. Test 2 neural network story drifts

Story	Case	Elastic drift	Inelastic drift
6	Earthquake -X	0.0011	0.0051
5	Earthquake -X	0.0012	0.0052
4	Earthquake -X	0.0011	0.0051
3	Earthquake -X	0.0010	0.0046
2	Earthquake -X	0.0008	0.0035
1	Earthquake -X	0.0003	0.0016
6	Earthquake -Y	0.0015	0.0067
5	Earthquake -Y	0.0016	0.0071
4	Earthquake -Y	0.0016	0.0071
3	Earthquake -Y	0.0015	0.0065
2	Earthquake -Y	0.0011	0.0050
1	Earthquake -Y	0.0005	0.0023

The maximum drift obtained was 0.0071, this value exceeds the maximum of 0.007 established by NTP E 0.30 for reinforced concrete buildings. However, the difference it exceeds is minimal and can be reduced by increasing its length by 5% or at the discretion of the designer.

3.2. Concrete saving calculation

The volume of concrete used in the building dimensioned with the use of neural networks and in a conventional way was obtained from the load measurements necessary for the seismic analysis, these volumes calculated for the neural network and the conventional method are shown in Table 6.

Table 6. Concrete volume comparison

Method	Volume of concrete (m^3)			Total volume (m^3)
	Column	*Beam*	*Shear wall*	
Neural networks	28.0	162.9	153.4	344.4
Conventional	21.8	129.1	232.2	383.1

It was observed that in the case of the building, a difference of 38.7 m3 of total concrete was obtained in favor of the building dimensioned with the use of neural networks. However, the volume of columns and beams is lower in the building analyzed conventionally, while the volume to be used in cut walls is considerably higher. This difference of concrete to be used in the first case represents a saving of 10.11% compared to the conventional method.

4. Conclusions

With the use of neural networks, a saving of 10.11% of used concrete was obtained in comparison to the same structure dimensioned in a conventional way. However, a greater volume of concrete was obtained in columns and beams due to the fact that there is a greater number of these structural elements, while in shear walls a notable difference was obtained by using a shorter length of shear walls in both directions.

The maximum drift obtained in the building dimensioned with neural networks is 0.0071, this value exceeds by 0.0001 the maximum allowed. However, this parameter can be reduced by increasing the length of the cut-off walls in a minimum or by placing them in a different shape, which concludes that neural networks can be used to obtain dimensions of structural elements that provide us with an optimal drift and consequently use a smaller amount of concrete for their construction.

The neural network elaborated in this research can be used in buildings that have a regular plan geometry and that the relation between the major and minor side does not exceed in 1.5, with approximate seismic parameters, a structural system of structural walls and of 6 levels with a mezzanine height of 3.00 m approximately. Since for the training stage, data with similar characteristics to those mentioned was used.

References

[1] B. Flyvbjerg, M. S. Holm, and S. Buhl, "Underestimating costs in public works projects: Error or lie?" J. Am. Plan. Assoc., vol. 68, no. 3, pp. 279–295, 2002, doi: 10.1080/01944360208976273.

[2] S. M. Nigdeli and G. Bekdaş, "Optimum design of RC continuous beams considering unfavourable live-load distributions," KSCE J. Civ. Eng., vol. 21, no. 4, pp. 1410–1416, May 2017, doi: 10.1007/s12205-016-2045-5.

[3] S. Chutani and J. Singh, "Design Optimization of Reinforced Concrete Beams," J. Inst. Eng. Ser. A, vol. 98, no. 4, pp. 429–435, Dec. 2017, doi: 10.1007/s40030-017-0232-0.

[4] J. Arcila Zea, C. Alberto Riveros Jerez, and J. Enrique Rivero Jerez, "Optimización de secciones de vigas y columnas para el cumplimiento de la deriva en edificaciones de concreto reforzado mediante Redes Neuronales Artificiales Optimization of beam and column sections for compliance drift of reinforced concrete buildings using Artificial Neural Networks," 2014.

[5] M. Kripka, G. F. Medeiros, and A. C. C. Lemonge, "Use of optimization for automatic grouping of beam cross-section dimensions in reinforced concrete building structures," Eng. Struct., vol. 99, pp. 311–318, Sep. 2015, doi: 10.1016/j.engstruct.2015.05.001.

[6] G. K. Bekas and G. E. Stavroulakis, "Machine learning and optimality in multi storey reinforced concrete frames," Infrastructures, vol. 2, no. 2, May 2017, doi: 10.3390/infrastructures2020006.

[7] M. Kamal and M. Inel, "Optimum Design of Reinforced Concrete Continuous Foundation Using Differential Evolution Algorithm," Arab. J. Sci. Eng., vol. 44, no. 10, pp. 8401–8415, Oct. 2019, doi: 10.1007/s13369-019-03889-5.

[8] C. A. Coello and A. D. Christiansen, "Multiobjective optimization of trusses using genetic algorithms." [Online]. Available: www.elsevier.com/locate/compstruc.

[9] A. L. Rojas, S. L. Chavarría, and M. M. Elizondo, "Optimization of reinforced concrete beams for rectangular sections with numerical experiments," Comput. y Sist., vol. 22, no. 2, pp. 599–606, 2018, doi: 10.13053/CyS-22-2-2542.

[10] A. Tomás-Espín and G. Sánchez-Olivares, "Optimum reinforcement and sizing of concrete cross sections subjected to flexural and axial loading 3305 Building Technology

[11] Blanco, A. (1994). Estructuración y diseño de edificaciones de concreto armado. Lima: Colegio de Ingenieros del Perú.

Evaluation of the Influence of Different Grades of Reinforcing Steel on the Seismic Performance of Concrete reinforced Frame Structures with Nonlinear Static Analysis

D Navarro[1], R Valero[1], J Orihuela[1]

[1]School of Civil Engineering, Peruvian University of Applied Sciences (UPC), Lima 15023, Peru

Abstract. In this investigation, the elasto-plastic behavior and the seismic performance of concrete reinforced frame structures reinforced are evaluated by applying the Pushover method. This evaluation is done on several cases: with high ductility steel (Grade 40), conventional steel (Grade 60) and high strength steel (Grade 75). For the previous, the capacity curve graph obtained from the displacement coefficient method was used to measure the capacity of the structure. In addition, the performance of the structure for different levels of seismic design are evaluated with the resulting values of ductility and rigidity of each case. The results showed that reinforcing a structure with a Grade 40 reinforcing steel increases the energy dissipation capacity, and if reinforced with a Grade 75 reinforcing steel increases the strength capacity in the structure. Finally, the comparative result of the various cases are presented to demonstrate the influence of reinforcing steel on the plastic behavior of concrete reinforced frame structures.

1. Introduction

In the last 30 years, the disasters that have been happening due to strong seismic movements have left structures with severe damage and, in other cases, uninhabitable structures. Among the most devastating are the following: the earthquake in southern Peru of 2001 of magnitude 8.4, Sumatra-Andaman of 2004 of magnitude 9.1, the earthquake of 2007 Haiti of magnitude 7.0 and the one on the Pacific coast in the region of the 2011 Tohoku of magnitude 9.1 [1,2,3,4]. Despite the fact that these earthquakes occurred in different places, they all evidenced similar seismic activity.

With reference to the aforementioned in the previous paragraph, structures subjected to large magnitude seismic movements present problems of reduction of stiffness and resistance to shear, which causes an increase in the demand for ductility. This causes the columns at the base to fail before the beams and as a consequence the structures collapse [5]. Regarding the influence of the steel grade in the reinforced concrete structure for structures subjected to severe earthquakes, the columns on the ground floor must be capable of developing sufficient ductility since when its reach their elastic limit, the concrete stops working and the Inelastic deformation is assumed solely by the properties of the reinforcing steel. For this reason, it is necessary for structures to dissipate a large amount of energy, so it is better to carry out a structural design with high-strength steel instead of designing with high-strength concrete [6]. Also, since the reinforcing steel in the beams is in tension, the steel yields before the concrete reaches its maximum compressive stress; therefore, it is suggested to reinforce with high ductility steel [8,]. 9On the other hand, buildings with columns reinforced with steels higher than grade 60 provide greater resistance to lateral load and a lower repair cost, compared to conventional steel [5]. Likewise, the amount of longitudinal reinforcement in the columns, it can be decreased in direct proportion to an increase in the elastic limit of steel [7].

For nonlinear static analysis and evaluation of seismic performance, a fundamental reference is the ASCE 41-13 code. This analysis has shown that the structural response is more accurate with the displacement design method than with the force-based method, since inelastic deformation is the main cause of possible structural collapse [10].

In this sense, this research focuses on evaluating the influence of reinforcing steel on the capacity, ductility, and stiffness of buildings in the face of seismic forces with non-linear structural analysis and evaluating the performance of the structure for different levels of seismic design for a basic objective.

2. Investigation Methodology

The computational models are first evaluated with linear static and dynamic analysis and then nonlinear structural analysis is performed. An incremental triangular lateral load pattern is applied, which were defined by approximating the inertial forces that will occur in a seismic event according to the predominant mode of vibration of the structure. From all this analysis the graphs of the building's capacity curves are obtained. The evaluation of structural performance with the Coefficient Method is evaluated according to ASCE 41-13 considering 4 levels of seismic demand defined according to their probability of occurrence and return period for the basic performance objective level. The validation of the results is done with the ATC-40, FEMA 273, BERTERO and VISION 2000 standards. The characteristics of the structural models are described in Table 1.

3. Investigation Methodology

The computational models are first evaluated with linear static and dynamic analysis and then nonlinear structural analysis is performed. An incremental triangular lateral load pattern is applied, which were defined by approximating the inertial forces that will occur in a seismic event according to the predominant mode of vibration of the structure. From all this analysis the graphs of the building's capacity curves are obtained. The evaluation of structural performance with the Coefficient Method is evaluated according to ASCE 41-13 considering 4 levels of seismic demand defined according to their probability of occurrence and return period for the basic performance objective level. The validation of the results is done with the ATC-40, FEMA 273, BERTERO and VISION 2000 standards. The characteristics of the structural models are described in Table 1.

Table 1. Description of the case studies

Case	With reinforcing steel (Degree)		Number of floors	f'c (kg/cm^2)
	Columns	Beams		
1	40	40	4	280
2	60	60	4	280
3	75	75	4	280
4	75	40	4	280
5	75	75	4	210

4. Results

4.1. Evaluation of the capacity of the structure

Regardless of what grade of steel the structure is reinforced, the resistance and displacement capacity is the same up to a certain point in the linear range. However, when the structure reaches its yield point, the displacement and resistance begin to vary, thereby demonstrating that it is important to perform non-linear structural analysis to assess the capacity of the structure more accurately.

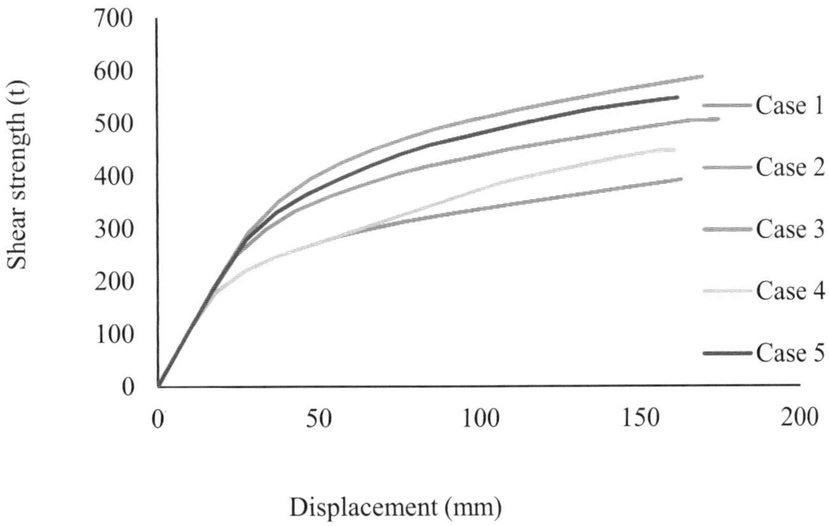

Figure. 1 Capacity curve of the 5 analysis cases

4.2. Evaluation of resistance capacity, stiffness and ductility

The conventional design of a structure reinforced with grade 60 reinforcing steel is 23% stronger, 29% less ductile and 2.3% more rigid compared to a structure reinforced with grade 40 steel. In contrast, it is 16% less resistant, 15% more ductile and 1.6% less rigid compared to a structure reinforced with grade 75 steel. This shows that the grade of the reinforcing steel influences the inelastic deformation of the structure.

The results of the investigation indicate that a structure with beams reinforced with grade 40 steel and columns with grade 75 steel is 16% less resistant, 28% more ductile and its rigidity is similar with respect to the structure reinforced with grade 60.

En cuanto a la estructura reforzada con acero de alta resistencia y con f´c=210kg/cm^2 es 7% menos resistente, y la ductilidad y rigidez son similares respecto a una estructura reforzada con el mismo grado de acero, pero con f´c =280 kg/cm^2. En cambio, es 8 % más resistente, 16 % menos dúctil y la rigidez es similar respecto a una estructura reforzada con acero de grado 60 y con f´c =280 kg/cm^2.

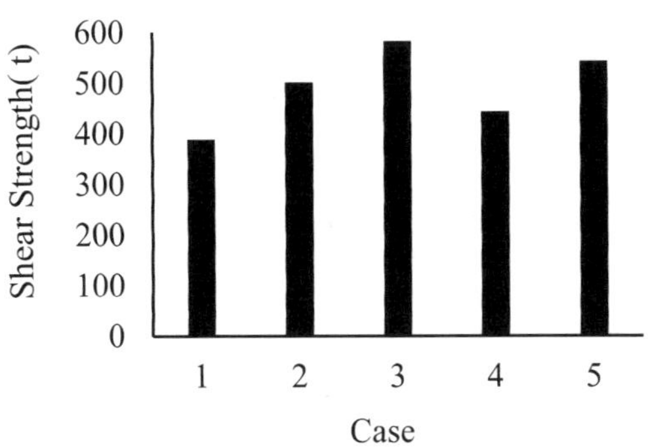

(a)

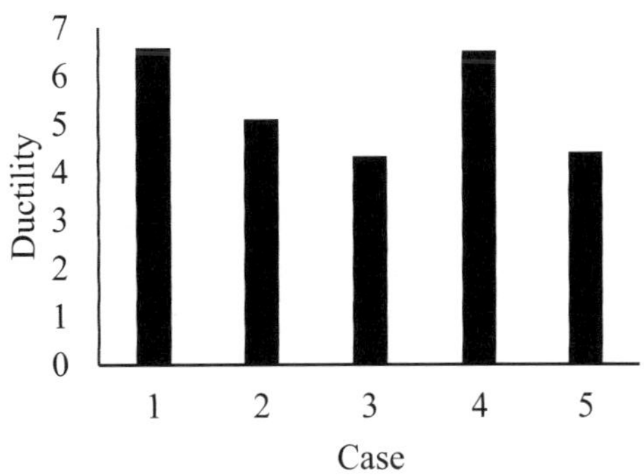

(b)

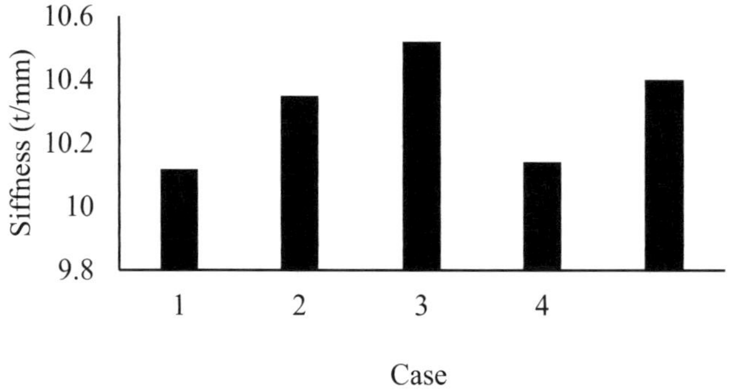

(c)

Figure. 2 Evaluation of resistance capacity, stiffness and ductility

4.3. Evaluation of seismic performance

The results indicate that the demand for resistance and ductility increases; however, the demand for stiffness decreases when the severity increases or the payback period is longer.

Buildings reinforced with high-strength steel (cases 3 and 5) tend to withstand higher strength demands and structures reinforced with high-ductility reinforcing steel in the beams and with high-strength reinforcing steel in the columns (case 4) it tends to resist the demand for stiffness more.

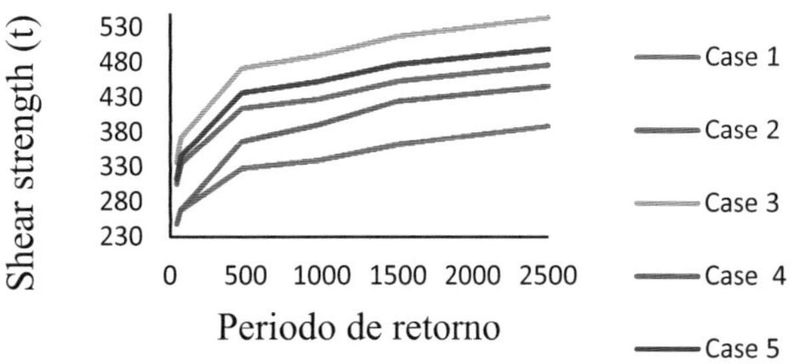

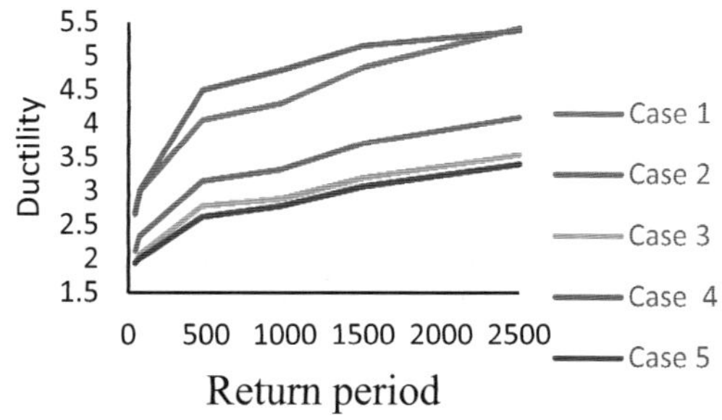

(a)

(b)

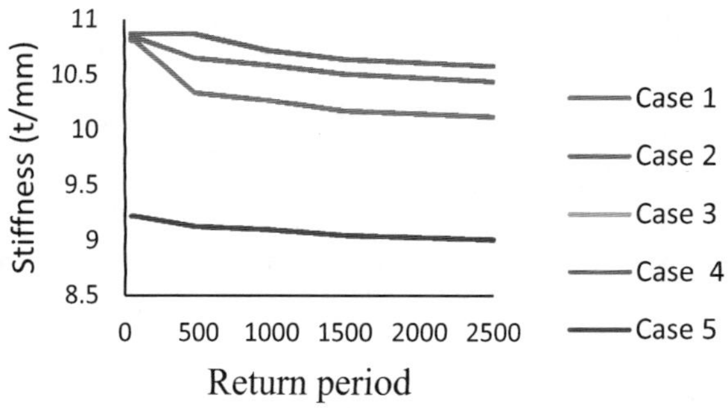

(c)

Figure. 3 Demand for shear strength, ductility and stiffness *a. Verification of lateral drift.*

The lateral drift presented by the structural models for a level of collapse prevention seismic performance indicates that they are lower than the limits proposed by the ATC-40, FEMA 273, and BERTERO and VISION 2000 standards.

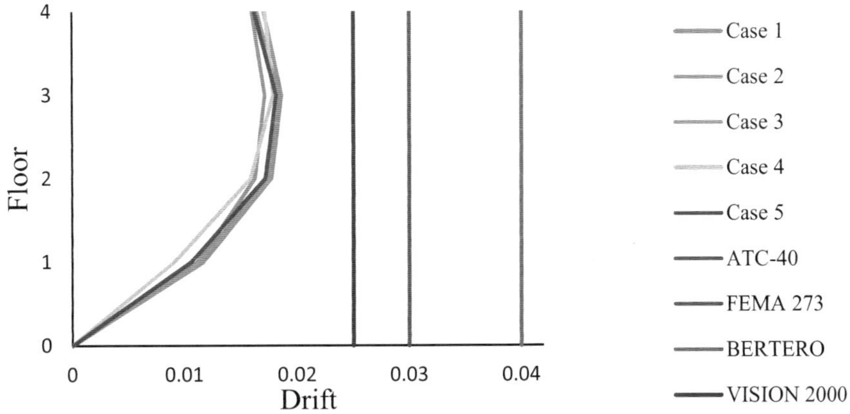

Figure. 4 Lateral drift demand and limits according to ATC-40, FEMA 273, BERTERO AND VISION 2000

5. Conclusions

Based on the results of the evaluation of the influence of reinforcing steel on the seismic performance of concrete reinforced frame structures, the following conclusions are made:

It is advisable to carry out the structural design combining different degrees of reinforcement in the structural elements, since each element supports different types of efforts, while some elements require more ductility, others require more resistance.

The beams that are subjected to traction must be reinforced with high ductility steel, since it provides the element with greater ductility. On the other hand, in the columns they need to be reinforced with high-strength steel in order to increase their resistance capacity to axial forces and lateral forces.

It is better to increase the resistance of steel than of concrete, because by reducing its f'c, the same ductility and resistance capacity of the structure is obtained.

For future research, the influence of reinforcing steel in tall reinforced concrete buildings can be studied with nonlinear static analysis.

References

[1] Giovanni M, Beck S and Wagner L 2018 *Geophys. Res. Let.* **29(21)** 1-14.
[2] Lay T, Kanamori H, Ammon C, Nettles M, Ward S, Aster R, Beck S, Bilek S, Brudzinski M, Butler R, DeShon H, Ekstrom G, Satake K and Sipkin S 2005 AAAS. **308(1)** 127-1132.
[3] Rochea R, Comercio M, Eberhard M, Mooney W & Rix G 2011 *Earthq. Spectra* .**27(1)** 1-21.
[4] Shibayama T, Esteban M, Nistor I, Takagi H, Nguyen T, Matsumaru R, Mikami T, Ohira K and Ohtani A 2012. *Works and Projects* **11(1)** 4-17.
[5] Ullad A, Lieping Y and Xinzheng L 2006 *Tsinghua Sci. Technol.* **11(6)** 640-647.
[6] Legeron F and Paultre P 2003 *J. Struct. Eng.* **29(2)** 241-252.
[7] Rautenberg J, Pujol S, Tavallali H and Lepage A 2011 *J. Struct. Eng.* **37(1)** 135-142.
[8] Jeffrey M, Pujol S, Tavallali H and Lepage A 2013 *ACI Struct. J.* **110(2)** 307-317.
[9] Puranam A and Pujol 2019 *ACI Struct. J.* **16(5)** 201-212.
[10] Mwafi A and Elashi A 2002 *Struct. Des. Tall Build.* **11(1)** 329-351.

Seismic Response of an Educational Building under Moving Mass Due to Earthquake Evacuation

C Bueno[1], F Arroyo[1], and V Arana[2]

[1] Bachelor, Peruvian University of Applied Sciences UPC, Lima Perú
[2] Professor, Peruvian University of Applied Sciences UPC, Lima Perú

E-mail: Type the corresponding author's e-mail address here

Abstract. Evacuating the occupants of a building during a seismic event influences their structural response. This phenomenon has been evaluated on bridges, continuous beams and platforms, which show significant changes in the performance of the structures. This research evaluates different cases applied to a building, where it seeks to identify the range of influence on structural performance and parameters that can be used in structure design methodologies. The main results of the structure analysis are: 1) the variation of the mass of the occupants, and 2) the variation of the structural response (base shear, periods and maximum drifts) for each case presented.

1. Introduction

One of the world's major sources of investment is infrastructure. The Global Infrastructure Hub estimated that the current spending trend is $79 trillion and presents a necessary investment forecast of $94 trillion in electricity infrastructure, roads, telecommunications, railways, water, ports, and airports [1]. However, this investment may be affected by the occurrence of a natural disaster such as earthquakes.

Damage to structures can be increased by effects not contemplated in current design methods. The research evaluates the movement of the occupants and its influence on the structural response of a building during the occurrence of an earthquake. There are other case studies in which similar analyses were performed for beams, platforms and bridges where the moving load was shown to have major influence on dynamic responses to an seismic event in materials used in cable-stayed bridge [4], while the interaction of vertical seismic loads with the inertial loads of movements amplifies the effects of moving vehicles on a bridge by up to 40% [5].5 For the case of platforms, the results presented were that the displacement response is higher at a normal speed and that the number of pedestrians influences the vibration characteristic [6]. In the case of bridges it was concluded that with the increase in mass speed the amplitude of vibration also increased and, furthermore, this increase in speed produced slight displacements, in addition that the maximum vertical moving load of the bridge, caused by an earthquake, decreases rapidly when it moves away from the epicenter by 40% for 20 km and 75% for 40 km distance [7, 8].

In this context, it was found that the movement of moving loads influences the response of the structure with speed, weight, eccentricity of the load and the material of which the structure is composed as important factors. In addition, it was show that seismic excitations can amplify the effects of mobile loads. This article quantifies the influence of the mass variation of the occupants on a building to estimate its effects. To do this, 15 cases of an educational building subjected to a moving mass have been modeled using the spectral modal analysis method.

2. Methodology

For the investigation, a compilation and survey of information is collected from the plans of a structure for educational use and data on the maximum number of occupants of the building to have a full capacity analysis. We proceed with data processing, where mathematical modeling of the structure is elaborated in the ETABS software. A spectral modal seismic analysis is developed under the consideration of the location of the masses at different points to obtain the structural results. Finally, we proceed with the comparison between static and moving mass analysis with the respective validation of results.

2.1. Case scenarios

The analysis structure is a seven-story armed concrete building, which is used as a university pavilion, in which it seeks to identify how it interacts with the change of position of a moving mass. The structure measures 18 m wide, 54 m long and height of 24.15 m; with rectangular columns of 50x110 cm and T-columns of 40x130 cm; Beams of 40x80cm, a slab 20 cm thick and a staircase that will be modeled as slab 30cm thick reinforced concrete, the model is shown in Figure 1.

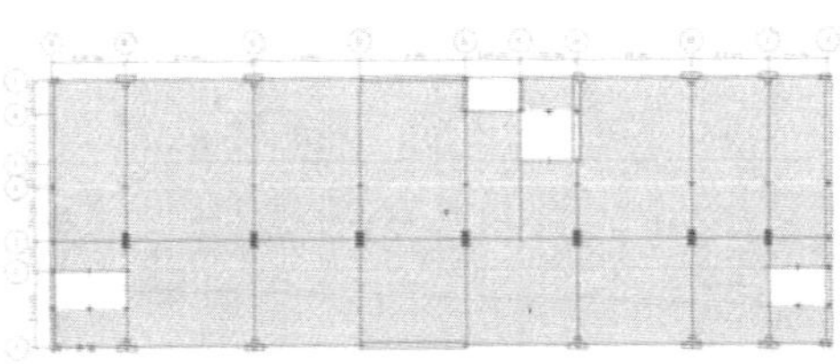
(a)

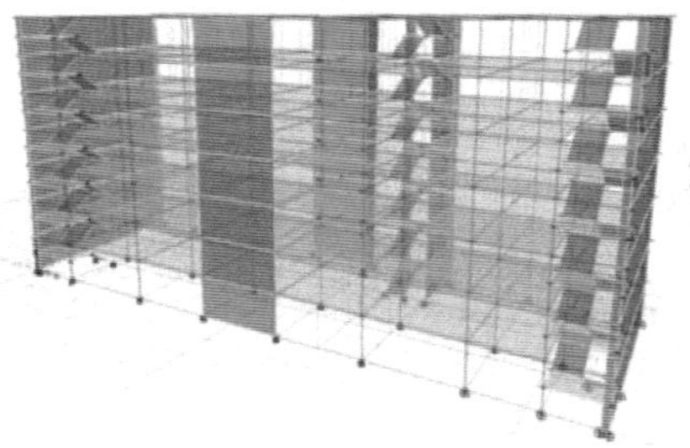
(b)

Figure. 1 Typical floor plan view (a), 3D model view of the structure (b)

2.2. Simulation of occupants in the building during evacuation

To analyze the influence of the moving mass of the occupants on the structure, different cases were raised to simulate this moving mass. A total of 15 cases were performed where the mass analysis was compared when the mass is static and located uniformly in the structure and when it changes position simulating the movement of the mass. The mass is represented as a living load distributed in the models of the structure.

In case 1, the live loads are distributed in the classrooms, bathrooms and hallways, case 2 was considered as a case of control since the loads are only in the classrooms, both cases were considered as static models where the applied load is the indicated by Peruvian regulations. Figure 2 presents the plan view of a typical floor with the distribution of loads in each case.

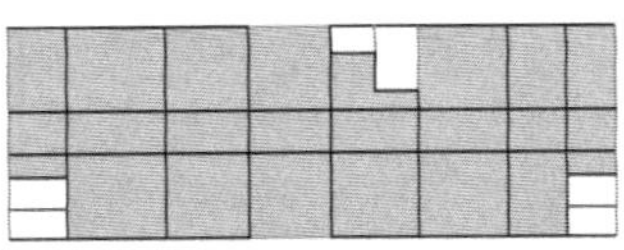
(a) Case 1

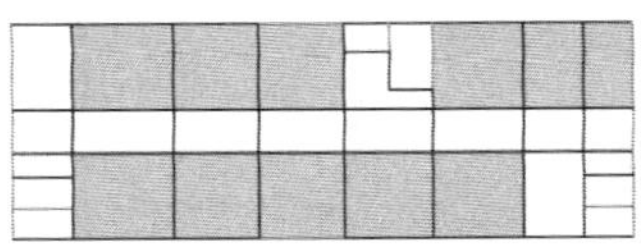
(b) Cases 2 and 3

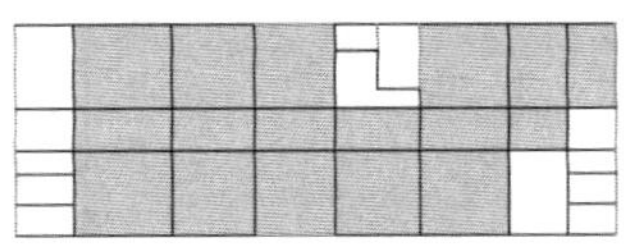
(c) Case 4

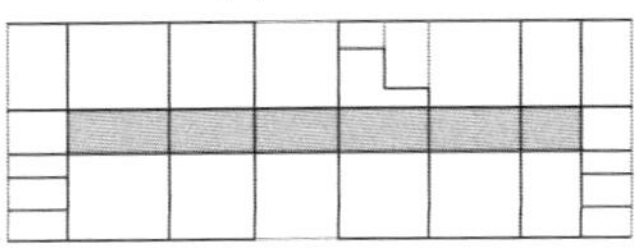
(d) Case 5

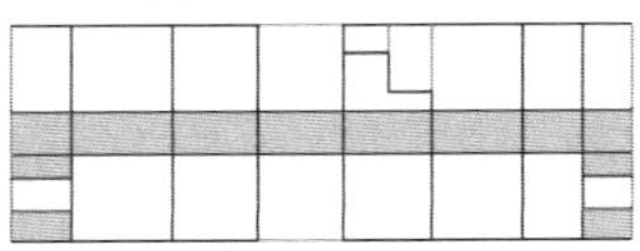
(e) Case 6

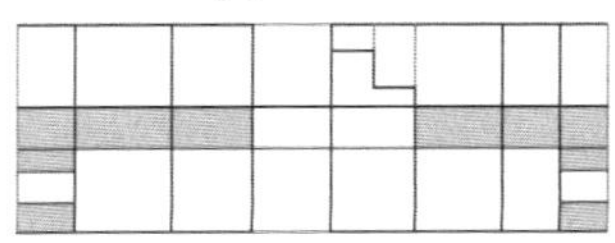
(f) Caso 7

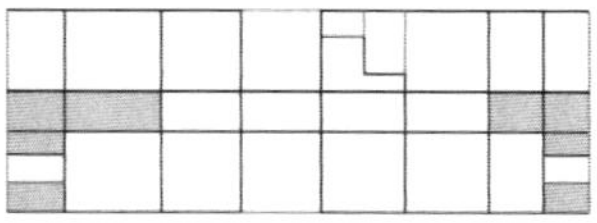

(g) Case 8

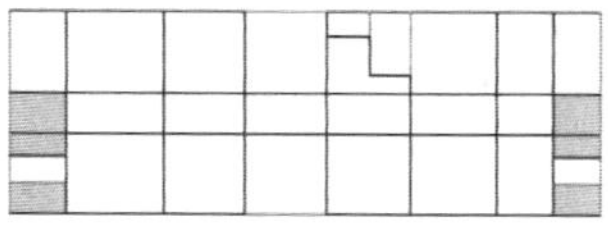

(h) Case 9

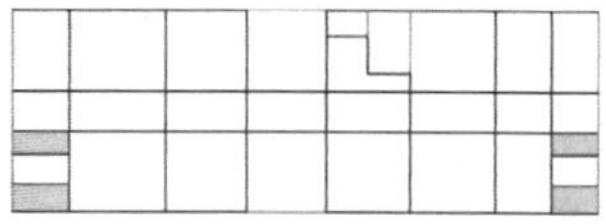

(i) Cases 10, 11, 12, 13 and 14

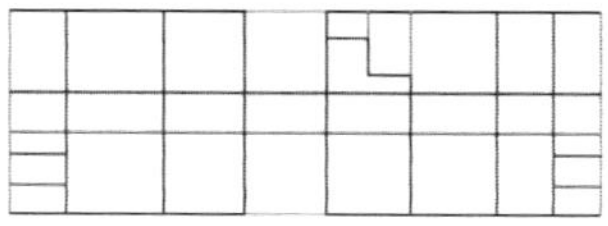

(j) Case 15

Figure. 2 Floor view of a typical floor of the distribution of loads in each case

From case 3, the scenarios where the mass is in motion are presented, in these cases an estimate of the weight of the furniture was used as fixed live load and an estimate of the actual weight of the occupants as a mobile live load, being this one that changes position. Figure 3 presents the front view of each case with their respective distribution of masses per floor.

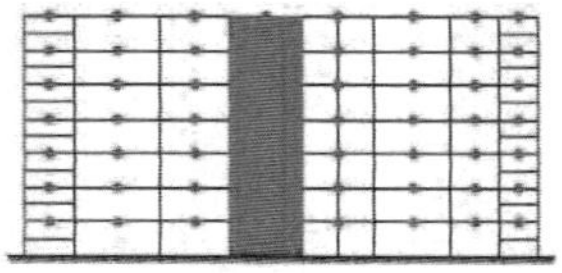

(a) Case 1

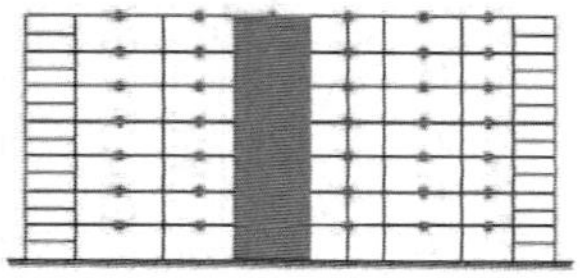

(b) Cases 2, 3, 4 and 5

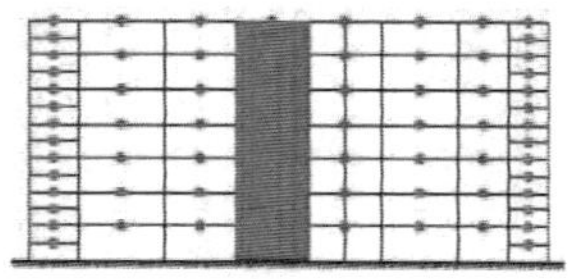

(c) Case 6

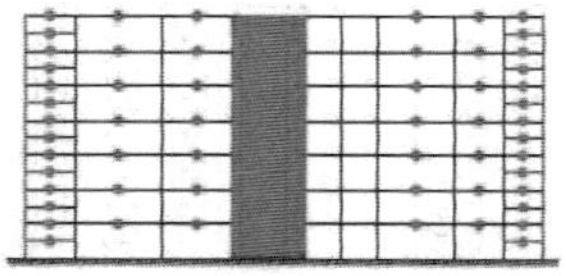

(d) Case 7

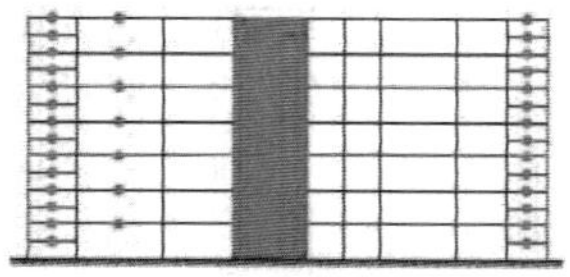

(e) Case 8

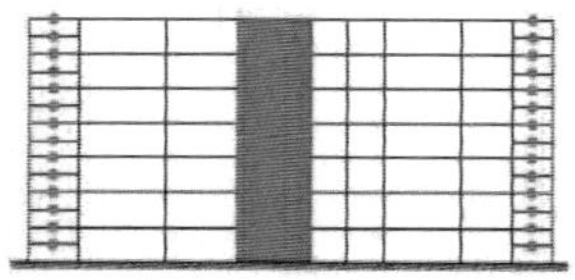

(f) Case 9, 10 and 11

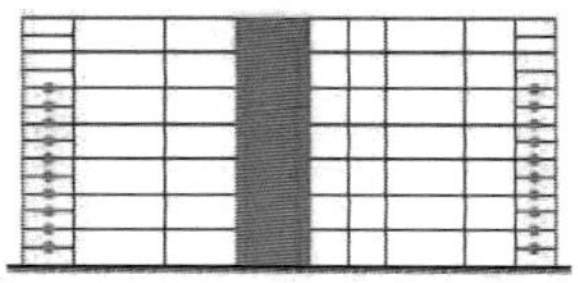

(g) Caso 12

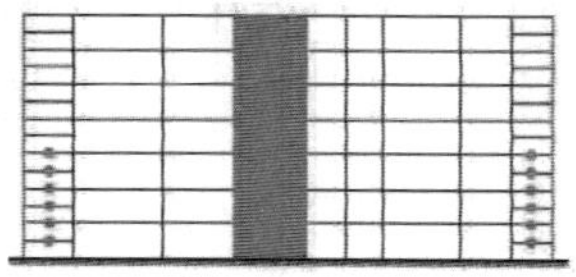

(h) Caso 13

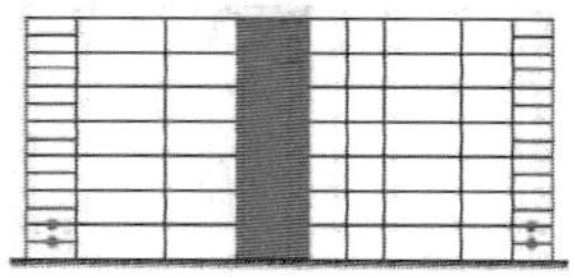

(i) Casos 14

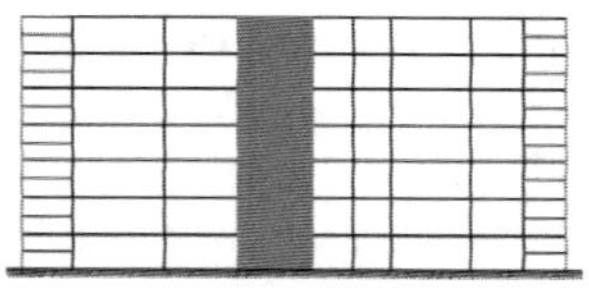

(j) Caso 15

Figure. 3 Front view of the load distribution on each floor on a case-by-case basis

For the model, a spectral modal seismic analysis is performed based on Peruvian earthquake-resistant design regulations. The structure is located in the city of Lima, in seismic zone 4 with an area factor of 0.45. The soil is Type 1, which is a rigid soil with a soil factor 1, as for the use factor, it is assigned the value of 1.5 since it is an essential building, the seismic amplification factor is 2.5. Finally, the load reduction factor for the X axis is 7, since according to the standard this axis corresponds to a Dual System, while on the Y axis the coefficient will have a value of 6 corresponding to a Structural Wall System. With the information collected, the respective acceleration response spectrum will be applied to each axis.

3. Results

3.1. Mass Variation

With the application of the loads in each case the masses and center of mass were obtained in the X and Y directions on each floor for each case analyzed. With the information collected, the total mass for each case was determined, these results are presented in Figures 4 and 5.

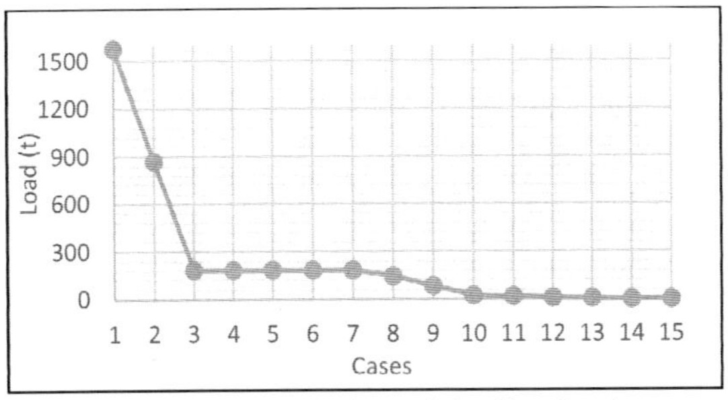

Figure. 4 Variation of the live load

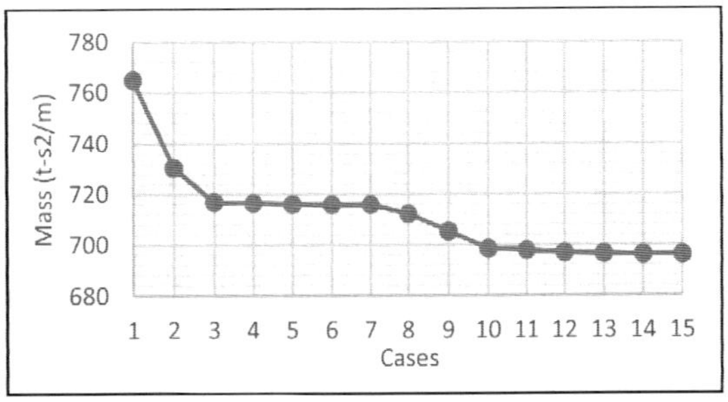

Figure. 5 Mass variation

3.2. Base Shear

To determine the base shear, a dynamic earthqueake was applies in the X and Y axes, the seismic parameters explained in the previous section were used in each, the shears were found for both earthquakes and the normalized base shear coefficiente, the results are shown in Figure6.

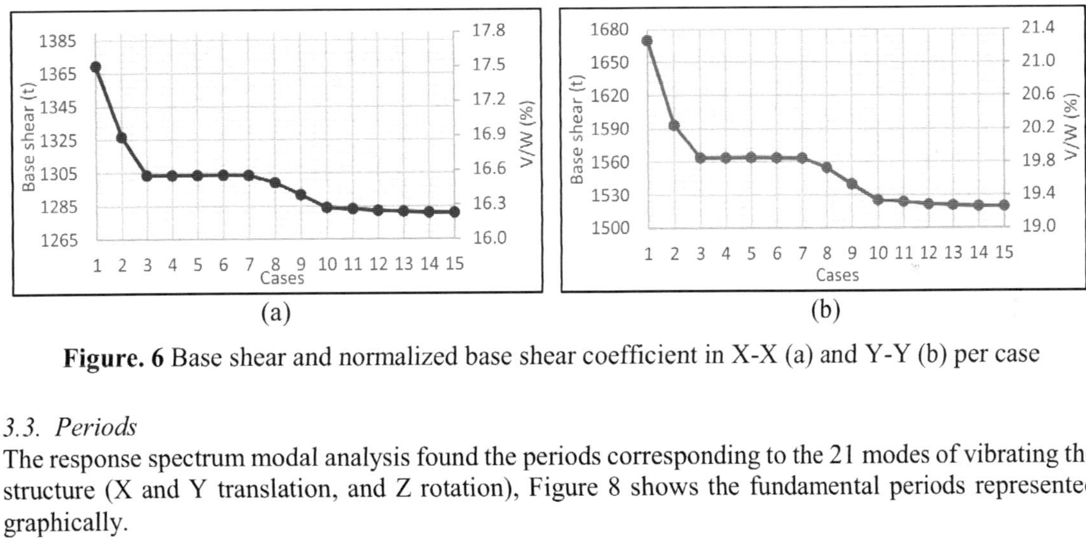

(a) (b)

Figure. 6 Base shear and normalized base shear coefficient in X-X (a) and Y-Y (b) per case

3.3. Periods

The response spectrum modal analysis found the periods corresponding to the 21 modes of vibrating the structure (X and Y translation, and Z rotation), Figure 8 shows the fundamental periods represented graphically.

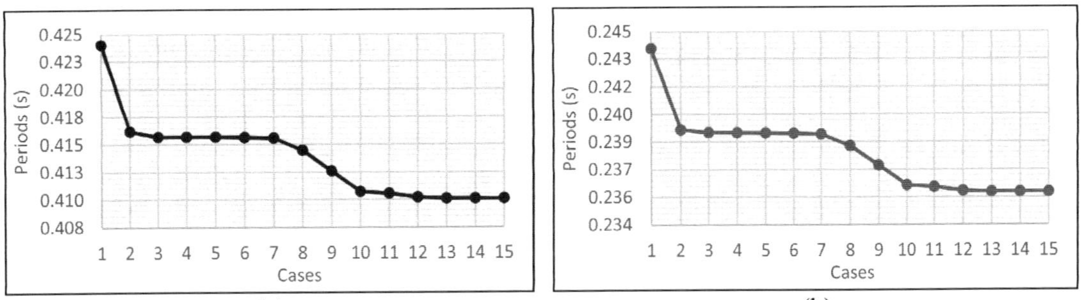

(a) (b)

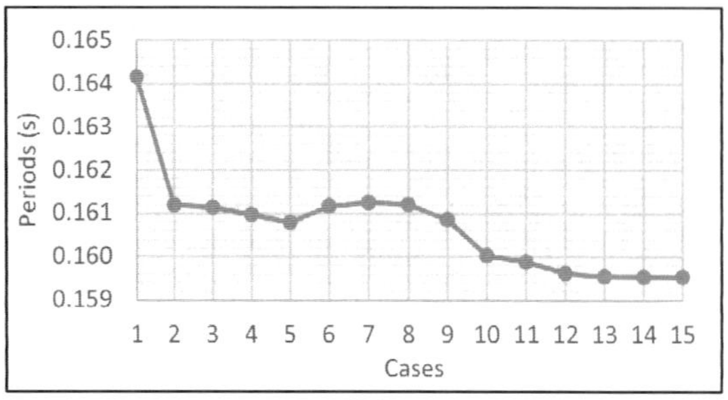

(c)

Figure. 7 Fundamental periods for each case in translation in X (a) and Y (b), and rotation in Z (c)

3.4. Drifts

For the case of drifts, inelastic drifts were obtained on each floor of the building and for each case raised. Having these results, the maximum inelastic drifts were found for both axes and graphs were made to easily visualize the variation that exists between each case, the results are shown in Figure 8.

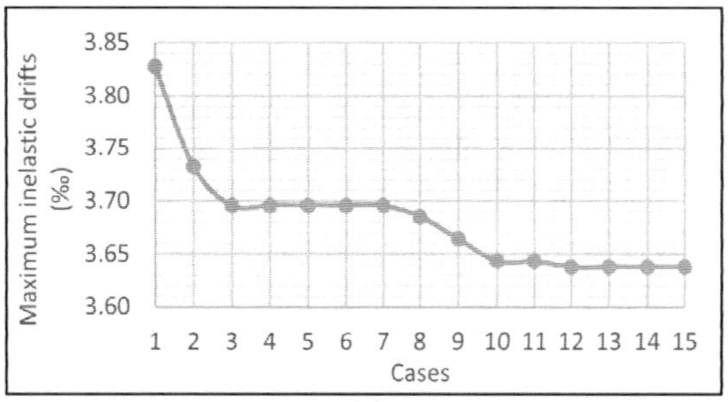

(a)

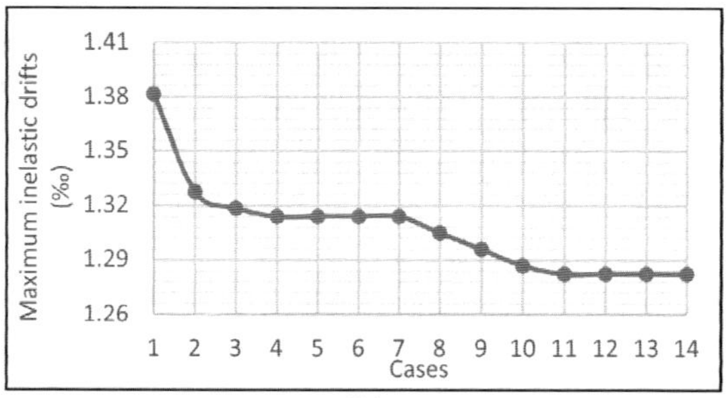

(b)

Figura 8. Maximum drift in X (a) and Y (b)

4. Discussion

It was observed that there is a gradual decrease in the live load between 1st and 6th floors, while on the 7th floor in the first two cases, corresponding to regulatory-based control cases, the mass is lower than the cases where the mass moves, this occurs since in cases of moving mass no load reduction or amplification factor is used, while in the first two only 50% of its total load is considered. As for the mass, it was found that although the load applied in cases where the actual load was estimated, the resulting mass on the top floor was 5.5%higher than that obtained in the case based on the standard where only 50% of the roof load is considered.

In relation to the base shear for dynamic earthquake on the X and Y axes, there is a higher shear on the Y axis with a value of 1670 t in the first case. The result obtained in the case of control is similar to those obtained in cases where the mass moves or changes position being in the range of 1590 t and 1520. On the X axis the shears for each case are in a range of 1370 t to 1300 t. As for the percentage of the weight of the building being displaced by seismic force the same trend was determined, in cases 1 and 2 this value was more than 20%, while in cases where if you simulate the moving mass this percentage was about 19%.

The results show that for case 1 the period in mode 1 is 0.424 s, 0.244 s in mode 2 and 0.164 s in mode 3, these values are higher than those obtained in the case of control with values of 0.416 s, 0.239 s and 0.161 for the first three modes, this is due to the fact that in the first case the analyzed building complies all the parameters of the standard, it also find loads in all areas. Making a comparison between case 2 and cases where the moving mass is observed that cases 3, 4, 5, 6, and 7 have periods similar to those of the control case, since in these cases there is no decrease in load on none of the levels. In addition, it was shown that the periods decreased as the mass decreased, these were affected by the movement of the mass. As for the periods corresponding to the rotation on the Z axis it was observed that in cases 6, 7, 8 and 9 there is an increase in its values, this due to the position of the load.

The maximum X-X axis drift value for case 1 is 0.003827, while in the case of control this is 0.0037, these values decrease in the following cases as the mass moves. The maximum drift on the Y-Y axis for case 1 is 0.00138, in case 2 it is 0.0013, in the following cases the drifts decrease as in those of the X axis. In the case of maximum drifts, it is obtained that the results of drifts in cases of moving mass decrease by 2.6% compared to the maximum drifts in the case of control, it is observed that the drifts are reduced due to the mass decreases floor by floor.

5. Conclusions

The models were made by changing the position of the loads to simulate the movement, scenario 2 was taken as a case of control and based on this it was observed that the results of the following scenarios do not show an increase in the structural response of the building, but decrease as soon as the mass decreases. However, a slight increase was observed in the periods corresponding to cases 6, 7, 8 and 9 in the rotation on the Z axis, this is because the mass is concentrated in the corridors and goes down the stairs, generating a large load at each end of the structure.

This increase in periods should be analyzed in detail, so that the effects can have not only on this structure, but also on irregular or high-rise structures. The methodology should also be improved to obtain a better analysis of the results, it is recommended to perform a history time analysis and use a program that allows to perform the analysis of the moving mass.

Referencias

[1] Oxford Economics. (2017). Global Infrastructure Outlook Infrastructure investment need in the Compact with Africa countries.

[2] Liu, M. F., Chang, T. P., & Zeng, D. Y. (2011). The interactive vibration behavior in a suspension bridge system under moving vehicle loads and vertical seismic excitations. *Applied Mathematical Modelling*, *35*(1), 398–411.Bazan E and Meli R 2002 *Diseño Sísmico de Edificios* (Mexico), Mexico

[3] Nikkhoo, A., Eskandari, A. B., Farazandeh, A., & Hajirasouliha, I. (2019). Vibration control of bridges under simultaneous effects of earthquake and moving loads using steel pipe dampers. JVC/Journal of Vibration and Control, 25 (19–20), 2580–2594.

[4] Ge, X., Wang, W., & Yuan, W. (2018). Research on dynamic characteristics of plate under pedestrian excitation based on Newmark-β. *Journal of Low Frequency Noise Vibration and Active Control*, *37*(4), 682–699.

[5] Chatterjee, A., & Vaidya, T. (2019). Experimental Investigations: Dynamic Analysis of a Beam Under the Moving Mass to Characterize the Crack Presence. Journal of Vibrational Engineering and Technologies, 7(3), 217–226.

[6] Konstantakopoulos, T. G., Raftoyiannis, I. G., & Michaltsos, G. T. (2012). Suspended bridges subjected to earthquake and moving loads. Engineering Structures, 45, 223–237.

AUTHOR INDEX